F I

The Institute of Biology's
Studies in Biology no. 52

Phytoplankton

A. D. Boney

D.Sc., Ph.D., F.I. Biol., F.R.S.E.
Professor of Botany, University of Glasgow

Edward Arnold

First published 1975
by Edward Arnold (Publishers) Limited,
41 Bedford Square, London WC1B 3DQ

Reprinted 1976
Reprinted 1979
Reprinted 1983

Paper edition   ISBN: 0 7131 2476 8

Printed by Photobooks (Bristol) Ltd.

General Preface to the Series

Because it is no longer possible for one textbook to cover the whole field of biology while remaining sufficiently up to date, the Institute of Biology proposed this series so that teachers and students can learn about significant developments. The enthusiastic acceptance of 'Studies in Biology' shows that the books are providing authoritative views of biological topics.

The features of the series include the attention given to methods, the selected list of books for further reading and, wherever possible, suggestions for practical work.

Readers' comments will be welcomed by the Education Officer of the Institute.

1983   Institute of Biology
                                                  20 Queensbury Place
                                                  London SW7 2DZ

Preface

If current editions of journals of biological abstracts are consulted the section dealing with algae will be found to contain a high proportion of papers on phytoplankton. The annual production of scientific work on plankton is immense, the underlying stimuli to research coming from fisheries, water supplies, cultivation of aquatic organisms and pollution studies. Directly or indirectly, all fish are dependent on the primary producers. In all parts of the world, in the sea or in lakes, some regions are highly productive and some very poor. The populations change in both numbers and composition with the passage of the seasons, and at each stage there is a complex pattern of interplay between chemical and physical factors and plant and animal populations. In this booklet I have set out to give account of the principal facets of current work on phytoplankton. This is a field of endeavour wherein many more questions are being asked than can possibly be answered at present. Both marine and freshwater habitats will be discussed, with emphasis mainly on coastal waters and lakes and less on rivers, estuaries and ponds. I hope that you will be stimulated to look at the changing populations of the floating plant life in a pond, small lake or region of inshore sea. This is still a field of scientific work where patient observation with the microscope can yield much new information.

Glasgow,   A.D.B.
1974

Contents

1 Introduction: The Organisms

Plankton is the living fraction of material which floats in the sea or freshwater and is moved passively by wind or current. It is composed of microscopic plants—the phytoplankton—which are predominantly autotrophic and are primary producers of organic matter in aquatic habitats. The nutritionally-dependent animal component constitutes the zooplankton. The phytoplankton thus stands on the base line of many food webs in aquatic environments and is in turn dependent on the activities of other microbial organisms, mainly bacteria, which convert organic material into the inorganic nutrients required by plants. Members of the phytoplankton are classed as algae.

1.1. The algae

The algae constitute a major grouping of plants in which representatives range in size and organization from microscopic one-celled organisms of a few micrometres diameter to highly organized macroscopic plants (e.g. the kelps) which in Pacific waters may attain lengths of 30 metres. Irrespective of size, the plant body of an alga is called a thallus since it lacks differentiation into stem, roots and leaves. All photosynthetic representatives of the algae contain chlorophyll *a* and, with the exception of the blue-green algae, this pigment is borne in distinct cell organelles called chloroplasts.

The pigment arrays in the chloroplasts give the observer the means of first grouping the algae into classes, but direct observation often has to be supplemented by chromatographic analysis. Colour differences in chloroplasts are due to the nature and quantities of the auxiliary pigments present in addition to the green chlorophylls. Thus in green algae the chloroplasts contain chlorophylls *a* and *b*, together with yellow-coloured carotene and xanthophyll pigments which do not mask the chlorophyll. Many members of the phytoplankton have chloroplasts which are brown, golden-brown or yellow, due to the green of the chlorophylls being masked by various xanthophyll pigments (Table I). All the pigments so far described are lipid-soluble compounds. Water soluble pigments of a proteinaceous nature, the phycobilins, are present in the blue-green algae and in some flagellates of the class Cryptophyceae. When present these phycobilins mask the chlorophylls and other pigments, giving the cells a blue-green or red colour depending on whether phycocyanin (blue) or phycoerythrin (red) predominates. Some unicellular planktonic algae are colourless and phagotrophic, i.e. the cells engulf and digest solid organic particles. The relationship between these colourless forms and other algae

1

Table 1 The distribution of pigments in algal classes. * = pigment present; (+) = pigment present in few genera; + ? = pigment present in trace quantities; – = pigment absent.

Pigments	Chlorophyceae	Prasinophyceae	Euglenophyceae	Chrysophyceae	Haptophyceae	Xanthophyceae**	Bacillariophyceae	Dinophyceae	Cyanophyceae	Cryptophyceae
Chlorophyll a	+	+	+	+	+	+	+	+	+	+
Chlorophyll b	+	+	+	–	–	–	–	–	–	–
Chlorophyll c	–	–	–	+	+	–	+	+	–	–
Chlorophyll e	–	–	–	–	–	+	–	–	–	–
α-Carotene	(+)	(+)	–	–	+	–	(+)	(+)	–	+
β-Carotene	+	+	+	+	+	+	+	+	+	–
γ-Carotene	+	(+)	–	–	–	–	–	–	–	–
ε-Carotene	–	–	–	–	–	–	(+)	–	–	–
Lutein	+	+	+	+	–	+	(+)	–	(+)	–
Zeaxanthin	+	+	?	–	–	–	–	–	+	+
Violaxanthin	+	+	–	–	–	+ ?	–	–	–	–
Neoxanthin	+	+	+	–	–	+ ?	–	–	–	–
Fucoxanthin	–	–	–	+	+	+	+	–	–	–
Diatoxanthin	–	–	–	–	–	–	+	+	–	+
Diadinoxanthin	–	–	–	–	–	–	+	+	–	–
Neodiadinoxanthin	–	–	–	–	–	–	–	+	–	–
Dinoxanthin	–	–	–	–	–	–	–	+	–	–
Neodinoxanthin	–	–	–	–	–	–	–	+	–	–
Peridinin	–	–	–	–	–	–	–	+	–	–
Neoperidinin	–	–	–	–	–	–	–	+	–	–
Myxoxanthin	–	–	–	–	–	–	–	–	+	–
Oscilloxanthin	–	–	–	–	–	–	–	–	+	–
C-Phycoerythrin	–	–	–	–	–	–	–	–	+	+
C-Phycocyanin	–	–	–	–	–	–	–	–	+	+

Phycoerythrin ⎫ found in
Phycocyanin ⎬ some genera

* Classes with planktonic representatives; ** = representative genera planktonic, but not extensively reported on in plankton studies.

has to be assessed on cell features other than chloroplast condition. The colourless algae will obviously be heterotrophic. Certain autotrophic pigmented algae, however, are known to exhibit phagotrophic properties.

1.2 Phytoplankton—general

In sea and lakes the diatoms and dinoflagellates (and in freshwater habitats the desmids) are the more obvious representatives of the phytoplankton, in terms both of cell size and availability, when water samples are examined under low-powered microscopes. We know very little as yet about the seasonal occurrences and contributions to primary productivity of the 'hidden flora' of very small microbial algae which, because of their size, are not often seen under the microscope and also escape capture by conventional methods of phytoplankton sampling. Generally these 'hidden flora' algae are green and brown flagellated organisms, and they sometimes undergo short-term population explosions called 'blooms' in response to local conditions. Some blooms of dinoflagellates give the 'red tide' phenomena observed in the inshore waters of the warmer seas. Certain blue-green algae also give rise to periodic summer blooms in freshwater lakes and in tropical seas. Cell size is clearly of importance with floating organisms. The largest known diatom is *Ethmodiscus rex* (diameter 2 mm). This species is usually found in tropical seas although it is also occasionally observed in the Atlantic. Table 2 gives a convenient scale for 'size grading' of phytoplankton organisms.

Table 2 Size grading of phytoplankton organisms

Maximum cell dimension	Plankton 'category'
1 More than 1 mm	Macroplankton
2 Less than 1 mm, retained by nets of mesh size 0.06 mm	Microplankton
3 5–60 micrometres (μm)	Nanoplankton
4 Less than 5 micrometres	Ultraplankton

The 'hidden flora' of the phytoplankton may be placed in the nanoplankton or ultraplankton category.

1.3 Recognition of phytoplankton organisms—what to look for

Most phytoplankton organisms are unicellular. The larger colonial forms possess individual cells that are usually of uniform structure and appearance. Some planktonic green and blue-green algae are of filamentous organization (i.e. form thread-like cell systems), and in some diatoms and dinoflagellates chains of loosely associated cells may be formed. Unicellular organisms are identified by recognizing certain cell characteristics, and ease of identification will clearly be limited by cell size and the degree of magnification obtainable. For identification purposes we look for the following with phytoplankton organisms:

(i) *Cell shape*   Is this constant or variable? The degree of cell flexibility is a function of cell-wall rigidity. Cells of the most variable shape often lack a rigid cell wall.

(ii) *Cell dimensions*   Measured in micrometres (μm = 0.001 mm) and requiring use of an eyepiece grid and stage micrometer.

(iii) *Cell wall*   Present or absent? When present, a cell wall is usually recognizable, although examination under high magnification may be necessary. The non-living components of the wall (e.g. cellulose and pectin) are detectable with suitable stains. Mineralization of the cell wall also occurs in many algae. In diatoms the wall consists of silica (SiO_2), and has complex surface patterns. Mineralized scales (SiO_2 and $Ca\ CO_3$) are seen on some marine and freshwater flagellates. Surface scales of organic material, often with elaborate designs, are seen on some microbial algae under electron microscopy. In the absence of a wall the cell is bounded by its periplast, a plasma membrane which may also be modified to form a pellicle; cell shape is highly variable. When the outermost bounding layer is a flexible non-living structure not in close association with the plasma membrane it is called a theca.

(iv) *Mucilage layers*   These are in fact extensions of the cell-wall system and are often not readily visible by direct examination. When organisms are mounted in weak indian-ink suspension the mucilage layers stand out as distinct haloes around the cells (Fig. 1–1).

(v) *Chloroplasts*   The three principal features to be noted are colour, number and shape. Colour is the expression of the pigment array present, and the shape and number of organelles are the means by which the organism presents a large surface area for light absorption and photosynthesis. Frequently a red-pigmented light-sensitive spot, the 'eye spot' or 'stigma', is associated with the chloroplast. A spherical body lacking pigmentation may also be seen in the chloroplast. This is the pyrenoid, which in green algae alone is associated with starch storage.

(vi) *Flagella*   Features of importance include the number of flagella, their points of insertion in the cell, relative lengths when more than one is present, the presence or absence of bristle-like outgrowths or of a covering of fine hairs, and presence of an additional flagellum-like structure called the haptonema (see p. 15), which is usually coiled and borne between the two larger flagella. The modes of vibration of flagella can also be of value in diagnostic work (Fig. 1–2a–d).

(vii) *Reserve substances*   Starch, oil and leucosin are the principal ones found in phytoplankton cells. Starch and leucosin are polysaccharides.

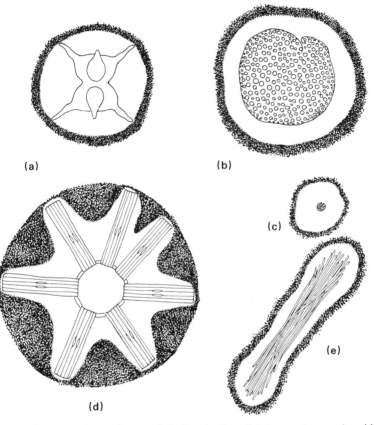

Fig. 1–1 Mucilaginous sheaths of planktonic algae. **(a)** *Staurastrum*, a desmid. **(b)** *Coelosphaerium*, a colonial blue-green alga. **(c)** Single cell of *Coelosphaerium*. **(d)** *Tabellaria*, a diatom. **(e)** Filaments of *Aphanizomenon flos-aquae*, a blue-green alga. Areas of stipple represent the indian-ink suspension.

Chemically these two are unrelated. Starch is a mixture of α-1,4-linked glucan and α-1,6-linked glucan (Fig. 1–3a and b), and leucosin is a β-1,3-linked glucan (Fig. 1–3c). Many plant starches have been described, but all would appear to be closely related. The starch reserve of green algae of the class Chlorophyceae is similar to that of higher green plants, giving the same blue-black colour reaction with iodine and producing the same compounds after chemical breakdown. Green algae of the class Prasinophyceae (see p. 13) have a starch reserve which gives a red-brown colouration with iodine. The Cyanophycean starch reserve of blue-green algae and the Floridean starch of the red algae are both closely related chemically to the starch of the Chlorophyceae. The euglenoid flagellates are also green algae, but

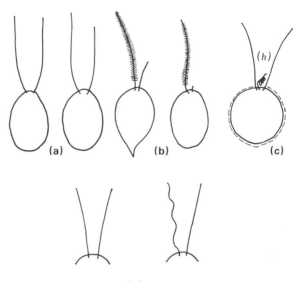

Fig. 1–2 Flagella of unicellular algae. **(a)** Left: smooth flagella of equal length; right: smooth flagella of unequal length. **(b)** Left: flagella of unequal length, one with hair covering and the other smooth; right: smooth flagellum almost a vestige. **(c)** Smooth flagella of equal length with coiled haptonema *(h)*. **(d)** Left: homodynamic flagella, vibration patterns similar; right: heterodynamic flagella, vibration patterns different.

their reserve substance, paramylon, is in fact a solid deposit of β-1,3-linked glucan. Leucosin is the main reserve of the brown or golden-brown members of the phytoplankton, and is a liquid stored outside the chloroplast in a distinct vesicle. It is probably the most abundant reserve substance occurring in unicellular planktonic organisms.

(viii) *Other cell features* In certain classes of organisms there will be additional points to be observed, e.g. cell vacuoles, small thread-like bodies ejected by the cell (trichocysts), whether the cell has a distinctive furrow or an apical intucking.

1.4 The diatoms (class Bacillariophyceae)

Diatom cells are ubiquitous, and are also unique in having a rigid silica-impregnated cell wall (the frustule) consisting of two parts (valves), with one (epitheca) overlapping the other (hypotheca) in the girdle region (Fig. 1–4a). The valves are linked in the girdle region either by pectinaceous bands or by the presence of small teeth fringing the inner edge of one girdle. This siliceous box encloses a lining cytoplasm, vacuole and nucleus. Chloroplasts may be either lobed structures (1–2 per cell) or many discoid

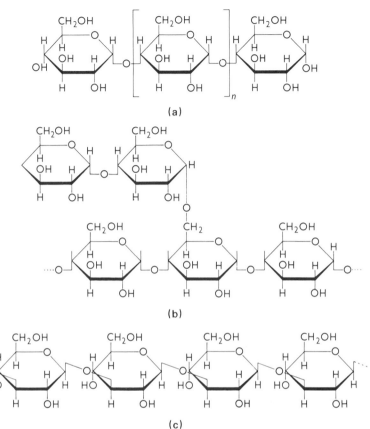

Fig. 1–3 **(a)** α-1,4-linked glucan (amylose). **(b)** Glucan with α-1,4 and α-1,6 linkages. **(c)** β-1,3-linked glucan.

bodies. Chloroplast colours are usually yellow-brown in planktonic species and dark brown in sessile and mud-living forms, these colours being due to the masking effects of xanthophyll pigments (e.g. fucoxanthin) over the chlorophylls present (Table 1). Reserve substances include lipid, volutin globules and leucosin in concentrated solution. The shapes of diatom cells can be complex, and are important in determining genera and species. Discoid (centric) diatoms can be likened to circular boxes and are radially symmetrical in valve view (circular or triangular, Fig. 1–4b) whereas the linear (pennate) diatoms show bilateral symmetry (Fig. 1–4c). Three planes of symmetry (valvar, apical and trans-apical) are seen with diatom cells. The surface patterning on the frustules is often very complex, consisting of pores which permit entry of nutrients and raised areas with regular patterning (puncta, costae and areolae). In some linear diatoms there is a slit-like

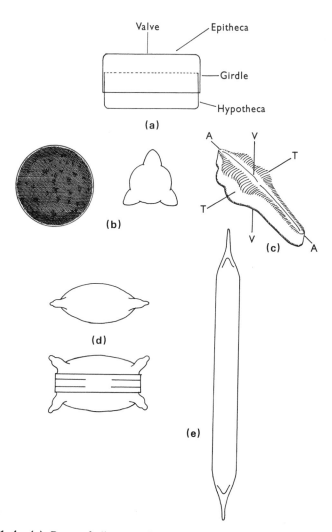

Fig. 1–4 **(a)** Parts of diatom cell—girdle view. **(b)** Left: discoid diatom (*Coscinodiscus grani*); right: *Lithodesmium*—both valve views. **(c)** Pennate diatom: A-A apical axis, T-T transapical axis, V-V valvar axis. **(d)** *Biddulphia rhombus*—Valve view (upper) and girdle view. **(e)** *Rhizosolenia*—elongate tubular cell. (**c** Redrawn after CHRISTENSEN, 1966.)

raphe on the valve surface with central and polar nodules flanked by regular lateral markings. The raphe allows contact between the medium and the cytoplasm. The pseudoraphe, a clear axial area bordered by regular lateral markings, is found in some linear diatoms. Linear diatoms may be wedge-shaped (cuneate), boat-shaped (corymbiform) or keel-like (carinoid). With some discoid diatoms the valve diameter may be appreciably greater than the axis length. Elongation of the axis leads to the gonioid form (e.g. *Biddulphia* species, Fig. 1–4d), and with solenoid forms there is pronounced development of the girdle substance with the formation of a tubular cell (e.g. *Rhizosolenia* spp., Fig. 1–4e); girdle length may be as much as fifty times valve diameter.

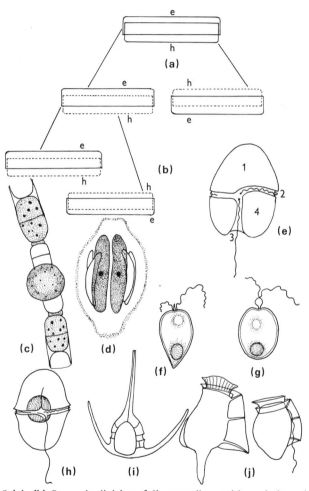

Fig. 1–5 (a)–(b) Stages in division of diatom cell: e, epitheca, h, hypotheca. **(c)** Auxospore formation in *Melosira borreri*. **(d)** Auxospore formation following conjugation in *Cymbella*. **(e)** Dinoflagellate cell organization (1 epicone, 2 girdle with transverse flagellum, 3 sulcus with longitudinal flagellum, 4 hypocone). **(f)** *Prorocentrum micans*. **(g)** *Exuviella marina*. **(h)** *Gymnodinium* sp. **(i)** *Ceratium tripos*. **(j)** *Dinophysis* sp. (**c** after GROSS, 1937; **d** after SMITH, 1950).

Plate 1 Scanning electron micrographs of diatoms (courtesy of F. E. Round, University of Bristol). **(a)** *Thalassiosira* sp. valve view × 1000; **(b)** *Stephanopyxis* sp. girdle view × 200; **(c)** *Cyclotella* sp. valve view × 1450; **(d)** *Skeletonema costatum* showing silica rods between cells × 1800.

The non-elastic siliceous wall maintains a constant cell size. In cell division the epitheca separates from the hypotheca, each forming the epithecae of the two daughter cells in which new hypothecae are formed (Fig. 1–5a). With continued and rapid cell division there will be a progressive diminution in cell size with some of the progeny (Fig. 1–5b), but with some species no such diminution is observed, indicating that some adjustment of frustule size must occur at the same time, probably in the region of the valve. When a progressive diminution does occur (to an extent that cells become too small to survive), this is halted in some species by the cell contents rounding off and forming auxospores. The auxospores, when released from the enclosing frustules, enlarge and form new siliceous walls and so restore the cell to its 'normal' size. Auxospore formation invariably results from gamete fusion (Fig. 1–5c,d). Colonies of cells are a feature of some plankton diatoms.

1.5   The dinoflagellates (class Dinophyceae)

These are widely distributed in marine, estuarine and freshwater environments. They are mostly unicellular and autotrophic, although some colourless parasitic forms are known. The cells bear paired flagella which arise in close proximity, usually with one (longitudinal) flagellum trailing behind the cell and lying in a groove (sulcus) and the ribbon-like transverse flagellum also lying in a groove (the cingulum or girdle) (Fig. 1–5e). The girdle lies between the epicone and the hypocone. Vibration of the longitudinal flagellum pushes water away, and the transverse flagellum causes cell rotation and forward movement. In a number of genera two flagella of unequal size arise from the cell apex (e.g. *Prorocentrum micans* and *Exuviella marina* (Fig. 1–5f,g). In addition to the biflagellate arrangement of longitudinal and transverse flagella described above there may also be modification of cell shape with flattening in an anterior-posterior direction, or from the sides. A thin wall (or theca) is present in some species (e.g. *Gymnodinium* sp. Fig. 1–5h), but other genera possess a theca of cellulose with a precise patterning of plates in some 'armoured' dinoflagellates (e.g. *Ceratium* sp. Fig. 1–5i). Wing-like extensions of the theca probably assist flotation in some genera (e.g. *Dinophysis* sp. Fig. 1–5j). The cells possess chloroplasts which are discoid or of various shapes and are yellow-green or yellow-brown in colour. The xanthophyll pigment peridinin has been identified in the chloroplasts of a number of species. Starch and oil are the principal reserve substances. The nucleus is large, with a characteristically beaded appearance. Other cell inclusions are the trichocysts and stigma. The trichocysts eject threads when stimulated.

1.6   Planktonic green algae (class Chlorophyceae)

The members of this class are commonly found in freshwater habitats but as a group they are far less well represented in the sea. Flagellum-

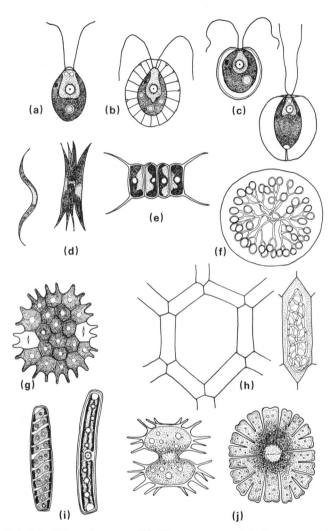

Fig. 1–6 (a) *Chlamydomonas.* **(b)** *Haematococcus.* **(c)** *Coccomonas* (upper) and *Pteromonas* (lower). **(d)** *Ankistrodesmis falcatus.* **(e)** *Scenedesmus quadricauda.* **(f)** *Dictyosphaerium pulchellum.* **(g)** *Pediastrum boryanum.* **(h)** *Hydrodictyon reticulatum.* **(i)** Saccoderm desmids—*Spirotaenia* (left) and *Roya* **(j)** Placoderm desmids—*Xanthidium* (left) and *Micrasterias.*

bearing cells occur as single-celled or colonial entities with flagella usually 2 or 4 per cell, of equal length, smooth, and arising apically. The chloroplasts are conspicuous, often single and cup-shaped (e.g. *Chlamydomonas*, Fig. 1–6a), with a prominent pyrenoid. Cell walls are of cellulose with a pectinaceous outer surface, but periplastic representatives are also known. In some genera the cell wall is a large envelope traversed by strands (e.g. *Haematococcus,* Fig. 1–6b) and in others a calcified envelope surrounds the cell (e.g. *Coccomonas* and *Pteromonas,* Fig. 1–6c). A prominent red stigma is associated with the chloroplast in many species. Osmoregulatory contractile vacuoles are present in most freshwater species. Certain non-motile planktonic green algae are of colonial habit, forming either loose associations of cells (*Ankistrodesmis,* Fig. 1–6d), or closely associated cells called coenobia (*Scenedesmus,* Fig. 1–6e). Colonial forms with cells lying in a loose mucilaginous matrix are typical of some freshwater species (e.g. *Dictyosphaerium,* Fig. 1–6f). The colonial organism *Pediastrum* consists of a flat plate of cells (Fig. 1–6g) and a network of large and regularly jointed cells is seen in *Hydrodictyon* (Fig. 1–6h). The desmids are a conspicuous component of the freshwater phytoplankton, and are mostly unicells. A small number of species are colonial in habit. The *saccoderm desmids* are usually oblong or rod-shaped cells with a single complete cell wall, and with chloroplasts of various shape (Fig. 1–6i). The *placoderm desmids* are characterized by great diversity of form, with cells of symmetrical halves (the semicells) joined at a narrow isthmus (Fig. 1–6j). The inner cell wall is of cellulose and the outer of pectin, frequently impregnated with iron. The cell wall may be extensively ornamented.

1.7   Planktonic green algae (class Prasinophyceae)

The representatives of this class are predominantly unicellular and are either bi- or quadriflagellate; some uniflagellate organisms have also been described. The flagella are thicker than those in the Chlorophyceae, and studies with the electron microscope have shown that this is due to scales of an organic nature which cover the surface of each flagellum, in two distinct layers in some species. The flagella usually arise from an apical depression (Fig. 1–7a). With many species organic scale coverings form the outer boundary of the cell; with others a theca (often flexible) is present (Fig. 1–7b). With some representatives a non-motile cyst phase alternates in nature with the flagellated cells (e.g. the genus *Halosphaera* common to north temperate seas (Fig. 1–7c).

1.8   The euglenoid flagellates (class Euglenophyceae)

These organisms are sparse in lake phytoplankton and are commonly found in ponds and temporary pools with high organic content. They are predominantly freshwater although common in estuaries. Both pigmented and colourless species are known. Flagella, usually two per cell, arise from an anterior invagination, usually with the long flagellum emerging and the shorter one fused with it in the region of the cell invagination (Fig. 1–7d). The emergent flagellum bears a row of fine hair-like outgrowths. Chloroplasts are numerous and of various shapes and the stigma is

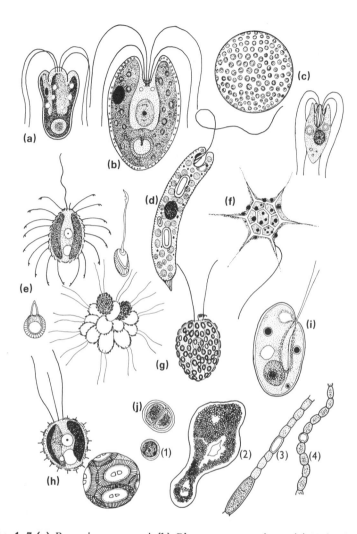

Fig. 1–7 (a) *Pyramimonas grossi.* **(b)** *Platymonas convolutae.* **(c)** *Halosphaera viridis*: cyst phase (upper) and flagellated cell (lower). **(d)** *Euglena* sp. **(e)** *Mallomonas* sp.: optical section (left) and scale (right); lower diagram: *Uroglena* colony (individual scale on left). **(f)** *Dictyocha speculum*—a silicoflagellate. **(g)** *Cricosphaera* sp.—cell with coccolith cover. **(h)** *Chrysochromulina* with organic scales and extended haptonema (upper) and *Coccolithus huxleyi*—cell with coccolith cover (lower). **(i)** *Cryptomonas* sp. **(j)** *Chroococcus turgidus* (1), *Microcystis* colony (2), *Aphanizomenon* filament (3), *Anabaena* filament (4). (**b** Redrawn after PARKE and MANTON, 1967; **d** redrawn after LEEDALE; 1967; **e** redrawn after BOURRELLY, 1970.)

a separate organelle. The reserve substance (paramylon) is present as a number of discrete bodies in the cells. The cell surface (pellicle) shows a helical (screw) symmetry with knob-like projections, and the organisms exhibit a characteristic and contractile form of movement.

1.9   The brown-coloured phytoflagellates

These unicellular organisms are characterized by having golden-brown or yellow-coloured chloroplasts due to the predominance of certain carotenoid pigments (e.g. fucoxanthin). They are mainly biflagellate, with apically inserted flagella, and store leucosin and oil as reserve substances.

CLASS CHRYSOPHYCEAE: biflagellate organisms with flagella of unequal length, the longer fringed with bristle-like outgrowths. Uniflagellate representatives are also known with only the long flagellum visible. Some genera bear mineralized siliceous scales of complex form (e.g. *Mallomonas* and *Synura*, Fig. 1–7e). These organisms are predominantly freshwater in habitat. The marine silicoflagellates, each cell with an internal skeleton of silica, are also members of this class (e.g. *Dictyocha*, Fig. 1–7f).

CLASS PRYMNESIOPHYCEAE: these are biflagellate unicellular organisms with smooth flagella of equal length. The cell contents are similar to those of the Chrysophyceae. With certain genera a third flagellum-like structure called the haptonema is present between the flagella, borne in a coiled or extended form. It was originally thought to be for attachment; but its function is at present uncertain. The genus *Chrysochromulina* (Fig. 1–7h) is found in freshwater and marine habitats, and is character-ized by possession of organic scales variously ornamented. In some genera these scales become impregnated with calcium carbonate, so forming the complicated patterns seen in the coccoliths (e.g. *Cricosphaera* and *Coccolithus* sp., Fig. 1–7g,h). The colonial form *Phaeocystis pouchettii*, which often forms conspicuous growths in coastal waters in late spring and early summer, has a motile flagellate phase with cell characteristics typical of this class (p. 79). (Class previously named Haptophyceae.)

1.10   Class Cryptophyceae

These are flagellated unicellular organisms with paired unequal flagella, both covered with fine hair-like outgrowths. The cells have a dorsi-ventral construction and are flattened in cross-section, frequently with a furrow passing over the anterior flagellar pole. The outer cell membrane is periplastic, with longitudinal striations. Chloroplast colours are variable, ranging from green to brown; red and blue-green chloroplasts are borne by certain genera, these colours being due to the phycobilin pigments phycoerythrin (red) and phycocyanin (blue). A cell intucking or gullet is present in many genera, and with certain colourless saprophytic forms this functions as a means of ingestion of organic particles (Fig. 1–7i).

1.11   The blue-green algae (class Cyanobacteria/Cyanophyceae)

The members of this class are distinguished from all other algae in being prokaryote, in common with bacteria, i.e. the cells are characterized by the absence of organized nuclei, lacking nuclear membranes and chromosomes; they are however capable of genetic replication. The cell components lack membranes, and the protoplasm is gel-like, without the streaming movements characteristic of eukaryotes (i.e. all other living organisms). The cell walls are distinct from those of other algae in consisting of two or three layers in close association with the plasma membrane. In some genera the chemical nature of the cell walls has been shown to be similar to that of certain bacteria. Planktonic blue-green algae are either unicellular, colonial or filamentous in habit (Fig. 1–7j). Both the cell colonies and the filaments cause extensive 'blooms' under certain marine and freshwater conditions. Filamentous blue-green algae possess specialized cells called heterocysts; these are the sites of nitrogen fixation. The cells of planktonic blue-green algae contain conspicuous gas vacuoles, presumably as aids to flotation.

1.12   Primary production

Plant life is the basis of all food webs in nature and hence constitutes the ultimate source of animal food. Most plants make this fundamental contribution by photosynthesis, utilizing radiant energy to synthesize from inorganic sources (carbon dioxide and water) organic compounds of high potential energy (PHILLIPSON, 1966). Chlorophyll a borne in the chloroplasts is essential to this complex synthetic process in these autotrophic organisms. Animals, bacteria and fungi which are dependent on performed organic substances for their food are called heterotrophs. Other than the seaweeds in marine habitats and certain aquatic flowering plants in lake margins, the phytoplankton is the principal source of organic material (primary production) in the sea and freshwater. Gross primary production refers to the total carbon fixed (or energy stored) by the phytoplankton. Net primary production is the amount of fixed carbon handed on to the first link in the food chain after the plants have used some of the organic matter for their own respiration.

All types of phytoplankton described for the different classes can make varying contributions to primary productivity depending on their abundance at the time of measurement. The contribution to energy turnover on the earth is considerable. Annual production of all plant life is estimated as 100×10^9 metric tons (tonnes) of fixed carbon. A major contribution to this organic matter production comes from marine phytoplankton. The oxygen liberated by phytoplankton photosynthesis is a vital part of the 'life support' system on the earth. Whilst production is measured as fixed carbon, the essential role of phytoplankton in food webs is to supply proteins, carbohydrates, fats, vitamins and mineral salts to primary consumers.

2 Factors Affecting Phytoplankton Growth

The growth of any plant requires light, carbon dioxide and water for photosynthesis, mineral nutrients in solution, and a suitable ambient temperature for metabolic activity. For the phytoplankton water is not limiting and the carbon dioxide supply rarely so, but the need to remain where there is sufficient light may well prove critical for an organism with a tendency to sink.

2.1 Light

The availability of sunlight as the source of radiant energy is an obvious feature of primary production. In aquatic habitats four aspects must be considered, viz. the means by which phytoplankton cells utilize this radiant energy; the intensity of the incident light; the immediate changes in the light on passing from air into water, and the extent to which with increasing depth this light both penetrates and undergoes further alteration. Illumination in all habitats will depend on the sun's position with latitude and season and on the cloud cover. In temperate regions the light intensity on a bright summer's day with a clear sky may be halved if clouds obscure the sun. By contrast, on a clear winter's day with bright sunlight the light intensity may be only one-fifth of that on a clear summer day, and with cloud cover this may be reduced to one-tenth. Hence there will be variations in light intensity over the course of a day, and in addition the wavelength composition will change through the day with movement of the sun. In north temperate regions the daily illumination reaches a maximum intensity in May and June, falling to approximately one-ninth of this summer level during December and January. In the tropics a daily illumination similar to that of the temperate summer lasts through the year except in the rainy season. In cold regions ice will allow light penetration into the underlying water, but snow cover reduces this. With freshwater reservoirs some control over the growth of phytoplankton is desirable and reduced illumination has been considered as a possible measure. Excessive algal growth will both clog the filter beds and affect the taste of the water. Various methods of artificial shading for control of plant growth have been suggested (e.g. use of plastic netting, or of aluminized plastic sheeting). Seasonal light intensity is closely associated with temperature although monthly changes in sea and lake temperatures are outpaced by variations in illumination.

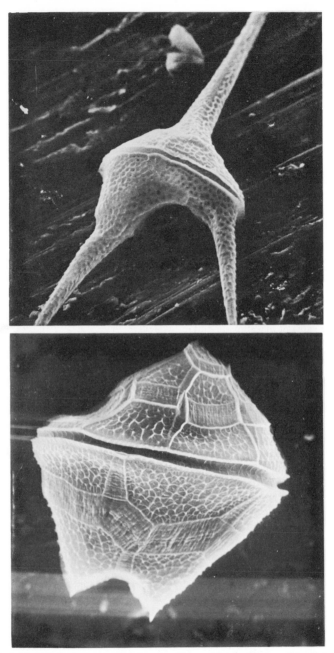

Plate 2 Scanning electron micrographs of two dinoflagellates, showing girdle regions and arrangements of thecal plates (x 1200). (a) *Ceratium hirundinella*; (b) *Peridinium leonis*. (Courtesy of J. D. Dodge, Birkbeck College, University of London.)

2.2   Chloroplasts

The chloroplast in a phytoplankton cell traps and utilizes light energy to convert carbon dioxide to carbohydrate and is the organelle system by which energy and carbon are obtained. Chloroplasts usually lie close to the cell membrane and show great diversity in size, shape, number and colour. Of the phytoplankton organisms only the blue-green algae lack discrete cell organelles of a chloroplast nature. In many phytoplankton organisms there are one or two large chloroplasts per cell, but in some (e.g. certain diatoms) numerous small discoid chloroplasts are present. The efficiency of the chloroplast as a light-trapping organelle is directly related to the surface area it presents to the incident light, and its effectiveness as a photosynthetic structure can be estimated by its output of carbonaceous compounds. Hence the variable shapes and parietal arrangements of the chloroplasts in planktonic algae, adaptations that ensure maximum light absorption and carbohydrate synthesis in relation to the volume of the cell. With phytoplankton organisms, as with all microbes, a large surface-area to volume ratio is an essential feature of their productive efficiency. The cell surface is the means of entry for both nutrients and dissolved gases and of exit for excretory products and extracellular substances. This allows the intensive rates of metabolism and rapid multiplication of phytoplankton cells shown in the seasonal population explosions which occur in seas and lakes.

2.2.1   Chloroplast pigments

With the exception of the blue-green algae the colours of other phytoplankton organisms are mainly due to the pigments in their chloroplasts. Chlorophyll a is the one pigment common to all and, in general, we may place phytoplankton in one of two major colour groups, those green- and those brown-coloured (p. 2). Two other chlorophylls, labelled b and c, are found in the phytoplankton, with chlorophyll b present only in those with green chloroplasts. Chlorophyll c has been identified in a number of brown-coloured organisms. Chlorophylls b and c differ from chlorophyll a both in molecular structure and in their absorption spectra in organic solvents. In the green and brown-coloured cells there are additional pigments, the carotenes and xanthophylls, and the relative predominance of these gives the brown and yellow-brown colours to some chloroplasts. In all cells, chlorophyll a is the primary pigment involved in photosynthesis and is the one pathway by which the absorbed radiation is converted to chemical energy; other pigments are secondary and must function on an auxiliary way, probably by absorbing radiant energy from certain wavelengths and transferring this to chlorophyll a. Some secondary pigments appear to be characteristic of particular classes. Chloroplast pigments can be readily extracted by means of organic solvents and separated by paper or thin-layer chromatography; they also show specific absorption spectra when eluted from the chromatograms. Pigment arrays are of value along with other cell features in ascertaining the class relationship of newly discovered organisms (Table 1).

The nature of the cell pigmentation is also useful in biochemical measurements of total phytoplankton biomass and in determining the classes of organism represented. When there are appreciable quantities of phytoplankton organisms in a natural medium, the cells can be filtered out, the pigments extracted, and the coloured solutions examined by spectrophotometry to obtain the relevant data.

2.3 Light intensity

In addition to the intensity of the incident light varying with locality, season and time of day, in water there is a gradual attentuation or decrease in light intensity with depth. This is due to absorption by the water and suspended particulate matter, (including the plankton organisms) and to reflection by the plankton and other suspended matter. There is some deflection of light by the water molecules (molecular scattering). Scattering of light by particulate matter also occurs; the blue colour of clear oceanic water is due to the upward scattering of blue light, whilst coastal waters appear green because the greater quantity of suspended particulate matter tends to reflect the light of longer wavelengths. In both the sea and in freshwater lakes three vertical zones may be envisaged with regard to available light. These are the euphotic (or photic), the dysphotic and the aphotic zones. The euphotic zone has enough light for photosynthesis but the dysphotic zone is too dim. The aphotic zone is dark and devoid of photosynthesizing plants. Hence any prolonged stay of phytoplankton cells in the aphotic or dysphotic zones may well prove harmful. The depths of these zones will depend on latitude and season and on local features such as proximity to land, the terrain drained by rivers and streams, and amounts of material in suspension. Hence it is impossible to define the depth of any one zone in absolute units. Measurements of light intensity at different depths have been reported either in energy units ($J cm^{-2} s^{-1}$) or in units of illumination (lx). Since both light intensity and its spectral composition can be limiting to phytoplankton photosynthesis and growth, seasonal measurements of underwater illumination will allow some assessments of the productivity of a locality. Underwater measurements with luxmeters are of little direct significance since these are instruments designed for use with normal daylight. They are more sensitive to certain wavelengths (e.g. green light) than to red or blue. Absorption in seawater filtered of all suspended matter is similar to that obtained in distilled water. Attenuation of the downward travelling light in water is due to a selective absorption and scattering of the wavelengths. The practical measure of this phenomenon for any one wavelength is made by determining the extinction coefficient, k;

$$k = \frac{2.3(\log_{10}I_{d_1} - \log_{10}I_{d_2})}{d_2 - d_1}$$

Where Id_1 and Id_2 are the light intensities at depths of d_1 and d_2 metres.

Red light is absorbed in the surface layers mainly by the water molecules. In coastal waters blue light is absorbed by suspended matter and by 'Kalle's yellow substance' which contains dissolved humic acids and is particularly found in the Baltic Sea. Hence where there is appreciable turbidity in the sea or lake waters the deep underwater illumination will be predominantly of green light (Fig. 2–1). Actual measurements of underwater radiant energy requires complex apparatus. A rapid

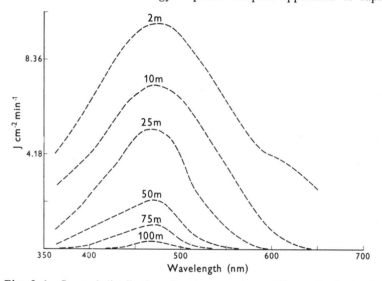

Fig. 2–1 Spectral distribution of radiant energy at different depths in clear ocean water. (Redrawn after JERLOV, 1951).

although subjective assessment of water transparency can be obtained with the Secchi disc. This is a white disc 30 cm in diameter which is lowered horizontally into the water (Fig. 2–2). The mean depth at which it disappears and reappears on being raised can be taken as a measure of the transparency of the water. It has been estimated that approximately 16% of the incident light at the surface penetrates to the depth at which the disc disappears in moderately clear water. The energy potential of this light will depend on the wavelength penetration (see above) (Fig. 2–3). Furthermore, whilst the depth of the euphotic zone will be variable with time of day and location, the level at which the light intensity is 1% of that at the surface

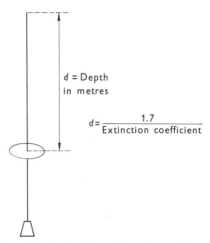

$d = $ Depth in metres

$$d = \frac{1.7}{\text{Extinction coefficient}}$$

Fig. 2–2 Secchi disc. The depth **(d)** estimated in this way is a rough approximation.

approximates to the lower limit of photosynthetic activity, and hence can be regarded as the physiological lower limit of the euphotic zone in both the sea and freshwater lakes. The absorption of radiant energy in passing down a water column is appreciable. In the sea, even in the cleanest ocean samples 62.3% of the incoming energy has been absorbed at a depth of 1 m, and 83.9% at 10 m. In coastal waters with much suspended particulate matter some 99.5% of the energy has been absorbed at 10 m depth.

2.3.1 Compensation depth

For any photosynthetic organism the production of organic matter by photosynthesis will be limited by the breakdown of carbohydrates for respiratory requirements. Where the available light is such that the rate of synthesis of organic compounds just keeps pace with the rate of respiratory breakdown then the compensation point is reached. With phytoplankton this point is a feature of depth, and will be the level at which the available light just permits photosynthesis to balance respiratory breakdown. At this depth the production of cell substance ceases (see Fig. 2–4). For temperate seas the compensation depth may be conveniently regarded as the level where the available light is 1% of that at the surface. This may be only 10 m in turbid inshore temperate waters compared with 120 m in clear tropical seas. Prolonged sinking of an organism below this level will place the cell under stress if the respiratory rate remains unchanged. In addition to the essential nature of underwater illumination, excessive light can be detrimental. Inhibition of photosynthesis has been observed with laboratory cultures of some phytoplankton organisms due to light saturation at high levels of illumination. Practical evidence that organisms from various depths have different light optima for photosynthesis has been ob-

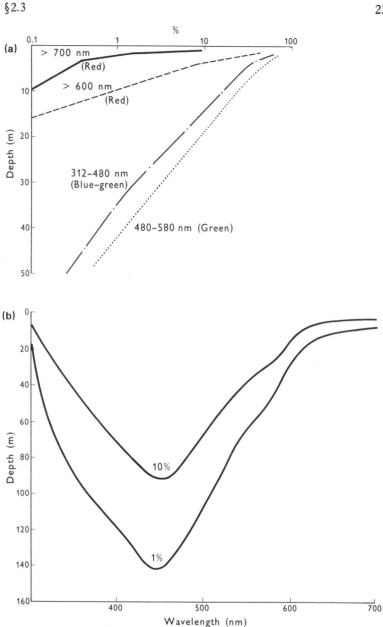

Fig. 2–3 **(a)** Change of intensity with depth of light falling on the surface of the sea (English Channel—autumn). **(b)** Depths at which the % of surface incident radiation reaches 10% and 1% for various wavelengths in clear ocean water. (**a** Redrawn after POOLE and ATKINS, 1937; **b** after JERLOV, 1951.)

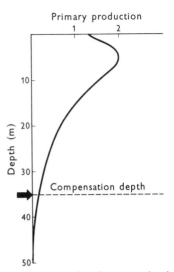

Fig. 2–4 Diagram illustrating change in primary production with depth down to compensation depth. Primary production measured in arbitrary units.

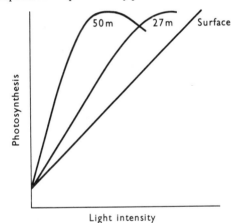

Fig. 2.5 Photosynthetic rates of Arctic phytoplankton, taken from different depths, with increasing light intensities. (After STEEMANN NIELSEN, 1952.)

tained in experiments with summer Arctic phytoplankton (Fig. 2–5), and continuous daylight with prolonged calm weather in Antarctic lakes has been shown to cause cell injury and severe inhibition of photosynthesis.

2.4 Temperature

We are concerned with two aspects; direct effects such as the temperature tolerance of organisms in relation to conditions in nature and the decreased solubility of oxygen in seawater with rise in temperature, and the indirect effects arising from temperature changes in the water mass. Laboratory experiments designed to test the temperature tolerance of phytoplankton organisms have sometimes given conflicting results. Thus a planktonic *Chlorella* species from Swedish Lappland isolated from a lake in which the temperature never rises above 7 °C was found in laboratory experiments to have an optimum of 20 °C. Measurements of temperature tolerances alone without precise simulation of other significant factors (e.g. nutrient supply) can lead to experimental results markedly at variance with field observations. There is a good deal of evidence that phytoplankton photosynthesis can proceed under temperature extremes, e.g. in Antarctic habitats below 0 °C and on tropical mudflats where temperatures may reach 30 °C or more. Some field observations suggest that certain organisms exhibit a seasonal pattern which is in part temperature controlled. Thus blooms of blue-green algae in freshwater lakes occur largely in summer (p. 79), and the summer abundance of dinoflagellates in temperate seas would seem to fit their relatively high temperature tolerances as shown in laboratory experiments. The species succession in the spring outburst in temperate seas has been described as temperature controlled, with the cold-tolerant species occurring in early spring and those favouring warmer conditions following in late spring and early summer. However, seasonal successions of plankton organisms are observed in both Arctic and Antarctic waters and in the tropical seas where variations in temperature are not wide. In the freshwater Lake Baikal in the USSR there is a population explosion of phytoplankton organisms before the ice breaks.

Whilst availability of carbon dioxide rarely proves limiting to plant growth in the sea and in freshwater habitats, in all aquatic habitats there

Table 3   Solubility of oxygen in water at different temperatures (cubic centimetres of oxygen at NTP dissolved in 1 dm^3 of water saturated with air at stated temperature and 760 mm pressure).

	Freshwater	*SeaWater* (*Salinity 35.4 parts per thousand*)
°C		
0	10.29	8.08
5	9.03	7.26
10	8.02	6.44
15	7.22	5.93
20	6.57	5.38
25	6.04	4.95
30	5.57	4.52

may be deficiencies in dissolved oxygen content due to temperature increases (Table 3). This deficiency, with its consequent effect on respiration, may prove limiting to plant growth.

The important indirect effects of temperature on phytoplankton organisms are to be seen in the thermal stratification of the water masses in both sea and lakes. This stratification is accompanied by formation of a thermocline or discontinuity layer (Fig. 2–6a). In the sea this requires a period of calm weather—which occurs most often in summer and accompanies a period of maximum air temperatures and insolation. In shallow seas with almost continuous turbulence the thermocline may never develop or last but a few days. In a land-locked water mass with minimal surface turbulence in summer, a clear-cut summer stratification is observed in which the upper layer of warmed water (epilimnion) is separated from the deeper cold water (hypolimnion) by a thin layer (metalimnion) in which there is marked temperature and density change (see Fig. 2.6b). Thermal stratification leads to isolation of the different water masses and of the suspended organisms, and when it occurs is a significant annual feature in the life of phytoplankton. An inverse temperature stratification in lake waters can occur in the late winter or early spring when the temperature of the top layers falls below that of the deeper water. In the sea the depth of the discontinuity layer is determined by surface turbulence but is usually a constant feature of low latitudes and is of seasonal occurrence in higher latitudes. Thus in the mouth of the English Channel in early summer a well-marked thermocline develops between 13–14 m depth in calm seas, whereas in neighbouring oceanic water it lies at 25–30 m. In the deeper water the waves are longer from crest to crest, and the surface turbulence due to their orbital movement extends deep down. In the shallow seas of the English Channel a summer gale may send the thermocline down to 18 m, and internal wave action may cause it to oscillate over a vertical height of 2 m or so (HARVEY, 1963).

Studies on Lake Windermere have produced much data on thermocline formation and stability. An example of the annual sequence is illustrated in Fig. 2–6c, which shows that in 1947 the establishment of a thermocline was measurable over a period of 16 days in May. Evidence has also been obtained from lakes that wind action on the surface can appreciably influence the stability of the thermocline. Wind across the lake causes a slight tilting of the surface, but the thermocline (and epilimnion) is moved down over a considerable distance at the leeward end of the lake whilst the hypolimnion thence slopes upwards at the windward end (Fig. 2–6d). When the wind action ceases, the thermocline tends to come back again to the equilibrium position. The swing back brings an increased momentum so that the tilting of the thermocline is now in the opposite direction. The overall momentum sets up a see-sawing action of the thermocline, pivoting about a central fulcrum and steadily decreasing in overall swing, which may last over a few days. Additional wind blowing across the lake may superimpose its own action and reaction on the original see-saw movement.

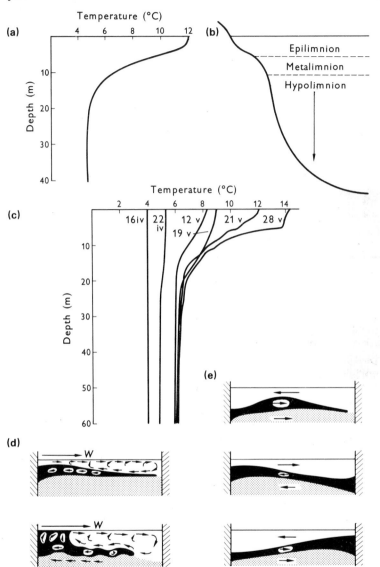

Fig. 2–6 **(a)** Temperature change with depth, showing thermocline formation. **(b)** 'Layering' of a lake in summer. **(c)** Thermocline formation in Lake Windermere. **(d),(e)** Seiche formation (diagrammatic)—**(d)** water circulation patterns due to wind action W (clear—epilimnion, black—metalimnion, stipple—hypolimnion); **(e)** oscillation of seiche. The arrows indicate the direction of water flow. (**c** Redrawn after MACAN, 1970.)

This oscillation of the thermocline sets up an internal seiche (Fig. 2–6e). Internal seiche formation is a feature of occasional wind action during summer months. Lakes in regions where there is protracted and frequent wind may lack thermocline formation during the summer, where orbital movements of the surface water are transmitted down through the water column and complete mixing occurs. Seiche formation has been measured for a number of freshwater lakes and periods of thermocline oscillation have been found to vary between a few hours, a few days (Loch Ness) and one month (Lake Baikal).

Similarly, frequent gales may prevent thermocline formation in open oceans (e.g. in the Roaring Forties of the south Atlantic). Turbulence of the water may be set in train by sea bed irregularities, and where these are big enough the disturbances may extend up to the surface layers. Upwelling of deeper water occurs when currents carry away the surface water. This phenomenon is particularly well shown in the coastal waters of West Africa and South America (Fig. 2–7). Such features tend to increase the mixing of surface and deeper nutrient-rich waters, producing dense growths of phytoplankton.

The significance of thermal stratification in terms of the organisms confined to the upper layers of sea and lakes lies in the barrier it sets up to vertical mixing. Above the thermocline, phytoplankton cells will be circulated by water movements and often carried into adequately lighted regions. Other factors limit growth. Replenishment of the plant nutrients in the surface layers depends largely on recycling processes of mineralization in the deeper waters and in the sea bed and lake floors. Minerals dissolved in land drainage are also significant but may often be insufficient for continued plant growth in the sea, where this terrestrial influence is mainly confined to coastal waters. Increased turbulence of the sea with the onset of autumnal gales sets in train the mass water movements which steadily lower and break up the discontinuity layer, so that vertical mixing will now occur. In temperate lakes the gradual fall in air and water temperatures leads to the autumnal phenomenon of 'overturn' and mixing.

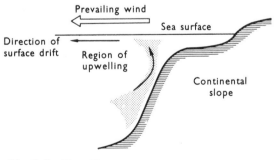

Fig. 2–7   Upwelling.

2.5 Plant nutrients

2.5.1 Nitrogen and phosphorus

Studies on plant nutrient requirements emphasize the importance of nitrogen and phosphorus. The first water-culture experiments with flowering plants conducted in the mid-nineteenth century by Sachs included nitrate, sulphate, phosphate and chloride as the principal anions, and potassium, calcium, magnesium and sodium as the main cations. A correct balance of six elements (nitrogen, phosphorus, potassium, calcium, magnesium and sodium) ensures the best growth of flowering plants. This evidence from use of artificial fertilizers with crop plants led to the early assumption that nitrogen and phosphorus would prove of great significance in influencing phytoplankton growth, and later measurements verified this assumption by showing the small quantities of each element actually present in natural waters. The importance of these two elements in both their effects on plant production and the seasonal changes in phytoplankton must not be underestimated, although deprivation of other elements present in trace quantities can also be limiting to plant production; this is probably true of certain organic compounds. Potassium appears to be of lesser significance with phytoplankton, a fact at marked variance with evidence from agricultural practices. The earliest culture media used with both freshwater and marine phytoplankton organisms were richly endowed with nitrogen and phosphorus (e.g. Miquel's solutions A and B, described in 1890).

Nitrate is an important nitrogen source for phytoplankton in both sea and freshwater. Other combined forms of this element do occur (ammonia, nitrite and organic compounds) and may well be utilized by some organisms in periods of nitrate starvation. Nitrogen in solution may be used by those blue-green algae capable of nitrogen fixation. Utilization of nitrate by phytoplankton involves its conversion ultimately to ammonia before assimilation into cell material, so it would seem likely that direct uptake of ammonium compounds would be advantageous. Whilst there is evidence from culture studies that ammonium N is used preferentially to nitrate, nitrate is present in much larger quantities in natural waters (Table 4). Associated with nitrate utilization are the observations that the enzyme nitrate reductase can be detected in the cells of some phytoplankton organisms and that this enzyme cannot be detected during periods of nitrate starvation and when other nitrogen sources (ammonia, nitrite) are available. The enzyme has in fact been detected freely in sea water from tropical regions but not shown to be directly associated with phytoplankton. Under conditions of protracted nitrogen starvation cells have been observed to accumulate fat at the expense of carbohydrates, possibly due to the enzymes associated with carbohydrate synthesis being more quickly influenced by the stress conditions than those linked with fat-synthesis.

Table 4 Plant nutrient concentrations in some natural waters. Data from Sykes J. B. and Boney, A. D., (1970), *J. mar. biol. Ass.,* **50**, 819; Heron J. (1961), *Limnol Oceanogr.* **6**, 338; Duthie, H. C., (1965), *J. Ecol.,* **53**, 361.

		Winter Maxima			Summer Minima		
		Phosphate P (μg at. dm^{-3})	Nitrate N (μg at. dm^{-3})	Silicate Si (μg at. dm^{-3})	Phosphate P (μg at. dm^{-3})	Nitrate N (μg at. dm^{-3})	Silicate Si (μg dm^{-3})
Marine	Menai Straits (Anglesey)	0.7–0.8	15	8–9	0.2	0.36	2.0
	English Channel	0.44–0.7	10.5	—	0.03–0.1	0.0	0.05–0.6
	Port Erin	0.67–0.69	5.8	6.7–6.9	—	—	—
	Cardigan Bay	0.80–0.96	25.2	8–12.0	0.0	0.36	0.17
		Phosphate P (mg dm^{-3})	Nitrate N (mg dm^{-3})	Silicate Si (mg dm^{-3})	Phosphate P (mg dm^{-3})	Nitrate N (mg dm^{-3})	Silicate Si (mg dm^{-3})
Freshwater	Windermere	5	0.4	1.2	0	0.1	0
	Llyn Ogwen	2.4	0.09	1.00	0.02	0.01	0.02

Filamentous blue-green algae (e.g. *Calothrix*) growing in salt-marsh and intertidal habitats make a significant contribution to the nitrogen budget by fixation of atmospheric nitrogen. This is generally a feature of heterocyst-bearing species. Many of the prominent blue-green representatives of the freshwater phytoplankton lack heterocysts, and whilst some planktonic heterocyst-bearing species also occur, certain of these (e.g. *Aphanizomenon flos-aquae*) do not appear to fix nitrogen. The view that nitrogen fixation does occur in freshwater lakes is based on indirect evidence from field observations, notably the very high concentrations of nitrogen associated with 'blooms' of blue-green algae compared with that available in combined form in the water, and measured uptake of ^{15}N. Of the marine planktonic blue-green algae, nitrogen-fixation has been attributed to the tropical genus *Trichodesmium,* but direct evidence seems to be lacking.

In natural waters phosphorus occurs in solution in both inorganic and organic forms; in lakes and the sea orthophosphates appear to be the main sources. Phytoplankton cells seem able to accumulate phosphate reserves well in excess of immediate requirements when nutrient levels are high—the so-called phase of 'luxury consumption'—and to utilize these reserves during periods of low phosphate concentration in the natural medium. These reserves enable cell growth to continue for some time after the level of nutrient in the water has been significantly reduced. Organic forms of phosphorus also occur in the sea and freshwater, and may serve as a source of the element for some phytoplankton during periods of deficiency. Enzymes able to break down organo–phosphorus compounds have been detected in lake waters, probably after autolysis of phytoplankton cells. With certain phytoplankton organisms in marine environments, phosphate deficiency has been accompanied by increased formation of the enzyme alkaline phosphatase, and production of this enzyme in the cells falls appreciably when phosphate levels in the external medium are again suitable for growth. Availability of organic phosphate compounds in the euphotic zone may be of appreciable ecological significance; utilization of such compounds would speed up recycling processes without need for total remineralization. The excretory products of zooplankton are another important source of recycled phosphorus.

2.5.2 Silicon

Diatoms require silica in soluble form for wall silicification (p. 7). The silicoflagellates are also dependent on silica for construction of their tubular skeletons, and some scale-bearing flagellates. At times of maximum diatom growth, natural waters show a decline in silica content. Evidence points to use of orthosilicic acid as the principal source of the element. The silica content of diatom cells expressed as a percentage of dry weight is considerable. With some planktonic freshwater diatoms it was found to vary between 26 and 63%, depending on species (LUND, 1965). Hence very low concentrations of silica in lake waters (0.5 mg dm^{-3}) have been considered as limiting to continued diatom growth, although some species (the marine *Skeletonema costatum*) seem able to grow with very thin siliceous walls. This organism appears in great numbers at the 'spring outburst' in some areas (e.g. the Clyde Sea and Southampton Water) and very thin walls would appear to be a likely consequence of this rapid cell proliferation. Utilization of silica by diatoms is linked with their sulphur metabolism, so that shortage of this element could indirectly prove limiting. Recycling of silica in the sea appears to be a fairly rapid process. The reappearance of silica in solution is more rapid after diatom frustules have been broken, as would happen during feeding by zooplankton. In freshwater habitats also the rate at which silica can be recycled significantly influences diatom periodicity. The rate at which the silica redissolves appears to vary with the species of diatom dominant. Where rates of solution are very slow replacement must come from inflowing tributary streams and rivers if diatom growth is to continue. Summer stratification in Lake Windermere results in the epilimnion being stripped of silica to below the critical level of 0.5 mg dm^{-3}. At such times a sudden rise in silica due to that brought down by inflowing streams can trigger rapid growth of the diatoms *Tabellaria flocculosa* var *asterionelloides* and *Fragilaria crotonensis* (LUND, 1964). Whilst it seems that silica depletion proves limiting to diatom growth in freshwater habitats, some authorities consider that in the sea this depletion is less significant because some recycling can take place.

2.5.3   Other mineral substances

Calcium appears to be present in sufficient concentrations for phytoplankton requirements in both sea and freshwaters. Certain desmids favour calcareous waters, but this is not thought to be due to any calcium requirement. The coccolithophorids require the element for formation of their calcareous scales, but there is no evidence of calcium deficiencies occurring in nature. Coccolith formation can be inhibited in culture in calcium-deficient media. Nor is there any clear-cut evidence that magnesium reaches growth-limiting concentrations in nature. Potassium seems rarely to be present in natural waters in concentrations so low that phytoplankton growth is inhibited; the small quantities of potassium compared with sodium in seawater has been considered due to the ease with which it becomes adsorbed on suspended particulate matter, with subsequent incorporation in bottom deposits. The importance of sulphur in the silicon metabolism of diatoms has already been mentioned; it seems to be present in sufficient quantities as sulphate in natural waters not to prove limiting to phytoplankton growth.

2.5.4   Trace elements (minor nutrients)

These are elements required by phytoplankton in very small quantities, but if present in insufficient supply they may limit phytoplankton growth. Most of the information concerning trace-element deficiencies and their effects on algae comes from laboratory studies. We lack data on their distribution and abundance in nature, and on whether they prove limiting to phytoplankton growth.

2.5.5 Iron

In freshwater and the sea iron occurs as particulate matter, in colloidal form, and in solution. The importance of the element to plants is that it is needed as a constituent of vital enzyme systems (such as cytochromes). The quantity in solution in natural waters is always very small, except under acid or reducing conditions in certain freshwater habitats. In the open sea a pH of 8 is usual, with minimal variation. More significant changes in pH occur in localized habitats (some inshore waters and rock pools), but these changes are rarely on the acid side. Hence in the sea the quantities of iron in solution will always be minimal, and are indeed so small that growth of phytoplankton could not take place without utilization of particulate and colloidal forms of iron. In 'brown water' lakes iron is present mainly in the ferrous state. The large quantities of organic matter present in some lakes may lead to complex ion formation with the element and may also bring about its reduction to the ferrous condition. There is experimental evidence that diatom cells may utilize particulate iron in contact with their siliceous walls, and some marine organisms have been found to assimilate insoluble ferric hydroxide under experimental conditions. Much of the iron in the sea is in the form of hydrous ferric oxide with particles of colloidal size. There is experimental evidence that this form of iron, which is known to adhere to cell walls, can also be directly used by planktonic algae. On the surface of plant cells a pH of 6 is likely, so that utilization at the surface would entail formation of soluble iron under conditions of little acidity. If the particulate iron is absorbed by the cells, lowered pH and reducing conditions would be available within the cytoplasm. C^{14} measurements of primary productivity have shown that lack of iron and other trace elements can prove limiting to photosynthesis of phytoplankton in the Sargasso Sea and north-west Atlantic. Similar evidence is available from parts of the Indian ocean near Australia. Coastal waters are usually more richly endowed with iron than the open sea. Sporadic 'blooms' of neritic diatoms in the open sea may indeed be induced by localized increases in iron or some other trace element through upwelling. There is some evidence that spring diatom outbursts are accompanied by depletion of iron in inshore waters (e.g. Puget Sound, USA, and the English Channel). From the results of laboratory experiments and from field data there seems little doubt that availability of iron significantly influences both the numbers and species composition of phytoplankton populations.

2.5.6 Manganese

Few species have so far been found to have a specific manganese requirement. The maganese content of coastal waters is higher than that of the open sea, and inflows are known to make a major contribution, particularly those from rivers flowing through rich arable land. As with iron, the quantity of manganese in solution is always very small, the particulate form being most common. Experimental studies have shown that enrichment of media with manganese will appreciably increase the growth of phytoplankton organisms. Whether shortage of the element proves limiting in natural waters is at present unknown.

2.5.7 Other minor nutrients

With chemically defined media in laboratory experiments a number of elements have been shown to be necessary for the growth of phytoplankton. These include copper, zinc, cobalt and molybdenum. There is at present insufficient evidence to show that in nature phytoplankton growth is limited by shortages of these elements.

2.5.8 Organic substances

The identifiable organic substances in natural waters include carbohydrates, amino-acids, fatty acids, organic acids and vitamins. In addition, plant growth substances with stimulatory or inhibitory properties have been reported. In ocean waters up to 18 freely occurring amino-acids have been recognized. Much of the organic matter comes from the decomposition of microscopic and macroscopic organisms, and from excretory products. Much soluble organic matter is also released into the water by healthy, actively growing phytoplankton cells in the form of extracellular products (p. 78). These extracellular products are not to be confused with organic substances released by damaged or dead cells.

Whilst there is experimental evidence from laboratory cultures that some phytoplankton organisms can show heterotrophic abilities with organic compounds, we lack conclusive evidence that photosynthetic respresentatives of the phytoplankton make much use of these compounds in nature. Sunlight remains the prime factor controlling seasonal change in algal populations.

More positive indications of the need for certain vitamins by algae have come from both laboratory and field observations. Three vitamins, vitamin B_{12} (also called cobalamine because of the presence of cobalt), vitamin B_1 (thiamine) and biotin, seem to be necessary. Bacteria are a source of these vitamins in nature but whilst some phytoplankton organisms are dependent on these sources, others seem able to synthesize the vitamins themselves (p. 64). Vitamin B_{12} and thiamine appear to be required more than biotin by photosynthetic algae unable to engulf particulate organic matter. Studies on vitamin B_{12} have been extensive, particularly with marine phytoplankton organisms. For some of these a definite B_{12} requirement has been demonstrated in culture experiments. Measurements of B_{12} content show that coastal waters are invariably richer than the open sea, and near the coast the concentrations (5–10 ng dm^{-3}) may be adequate for plant requirements at all times of the year. A seasonal cycle of B_{12} abundance has been indicated for the North Sea, with a winter maximum and summer minimum. There are indications from laboratory experiments that with certain diatoms the B_{12} requirement is of greater significance at some stage in their lives (e.g. auxospore formation). There is much less information on the B_{12} requirements of freshwater phytoplankton. Thiamine is also necessary for the growth of some marine phytoplankton organisms and, since it also is more abundant in coastal waters than in the open sea, lack of this vitamin might influence phytoplankton succession in the oceans.

The part played in nature by extracellular products of actively growing phytoplankton cells has attracted much attention. In Lake Windermere the quantities excreted have been estimated as equal to 50% of the fixed carbon during the period of the spring outburst and between 10–70% in the midsummer period when diatom numbers are small. Coastal phytoplankton organisms have been shown to excrete 35% of fixed carbon. Glycollic acid, one of the most studied of these extracellular products in natural waters, may enable heterotrophic growth of algae during winter months and support the growth of bacteria which in turn release the extracellular products of their metabolism. Appreciable quantities of fixed nitrogen are released by blue-green algae. Whilst these are products of some nutritional value, various substances described as growth promoters and growth inhibitors have been reported from natural waters. The production of organic substances by phytoplankton organisms and their utilization by others underlines what is now an accepted feature of primary production: that successive crops of organisms imprint on a water mass a biological history which must influence those which follow. Sometimes these influences are beneficial, and sometimes toxic (p. 79).

2.6   Eutrophic and oligotrophic lakes

This is a classification based on availability of plant nutrients. Eutrophic lakes have a good supply of nutrients and potentially a high productivity; oligotrophic lakes are the opposite. A eutrophic lake will support a rich phytoplankton flora and animal population. The process of eutrophication can be a long-term enrichment or ageing process in natural waters. Artificial (man-made) eutrophication will appreciably speed up this process, sometimes with dramatic side effects (p. 97). The rich nutrient supply in eutrophic lakes can result in dense growths of phytoplankton which significantly reduce light penetration. Hence a high rate of productivity is restricted to the upper layers. When measured in terms of organic production per unit of surface area the productivity of a eutrophic lake may resemble that of an oligotrophic lake. In an oligotrophic lake the less dense plankton will allow light penetration to greater depths. With the deeper photosynthetic zone, production per unit surface area of an oligotrophic lake can be similar to a densely populated eutrophic lake. But, because of tne lower nutrient levels, the productivity per volume of an oligotrophic lake will always be lower than that of one which is eutrophic. Another feature relevant at the consumer level is the ease of capture of food. A sparse algal population will entail greater energy expenditure by the predator. In dense phytoplankton more than enough food can be filtered with little movement on the part of the animal. The digestibility of the food is also important. Even with a dense phytoplankton population in a eutrophic lake it is possible that the larger diatoms will not be acceptable as food for zooplankton.

2.7 Salinity

Variations in the total salt content of water, ranging from the brackish water of estuaries to the high salinities of the open sea, will present barriers to the spatial distribution of phytoplankton organisms. Some phytoplankton organisms as a group are entirely restricted to fresh water (e.g. the desmids). Inflow of salt water into areas of fresh water will have catastrophic effects on the phytoplankton, and one of the principal causes of this is the sodium chloride of sea water (30.4 parts per thousand). Some organisms can survive and grow in the variable salinities of estuaries. With some nanoplankton representatives from estuaries, marked changes in volume can be measured in cells subjected to low salinities without damage to cell functions (e.g. in 5 parts by weight of total salts per thousand parts by weight of water compared with 35 parts by weight of salt in ocean water). Diatoms and dinoflagellates not adapted to withstand osmotic stress (e.g. not euryhaline in their properties) will clearly suffer cell damage if suddenly subjected to fresh water in the case of marine organisms, with cell distortion and rupture, or rapid plasmolysis as when fresh water phytoplankton pass into the sea. Large-scale studies of ocean phytoplankton show that salinity can rarely be dissociated from temperature in defining spacial distributions. Species typically found in the Antarctic ocean are probably adapted to low temperatures (−1.8 to 3.5 °C) and salinities ranging from 32.6 to 34.5 parts per thousand. By their restricted range of temperature and salinity tolerance they are described as stenothermal and stenohaline. In tropical seas where there are temperature fluctuations but again a restricted salinity range, certain organisms are eurythermal and stenohaline. Temperature and salinity fluctuations can be wide-ranging in estuaries, and organisms adapted to both types of stress are both eurythermal and euryhaline.

Barrage schemes have been proposed for a number of estuaries in the British Isles. If effected, these will ensure improved water supplies since the enclosure of an estuary will result in the build-up of fresh water inflowing from a sizeable catchment area. The transitional stages of development will be interesting in terms of phytoplankton successions.

2.8 Laboratory culture techniques

For any conclusive study on nutrient requirements, or on the release of extracellular products, cultures of algae freed from bacterial contaminants

are obligatory. There is now a large number of culture media, each formulated in terms of nutrient enrichments of the sea and lakes. In some
cicumstances entirely artificial media are prepared using glass-distilled
water, and in others an aged natural medium is enriched. The formulae of
some of the commonly used culture media are listed in the Appendix. Conditions of light and temperature are also important. Cultures set up with
'north window' illumination are often favoured to avoid excessive irradiation of the organisms and heating of the media. Room temperatures of
about 18–20 °C can be tolerated by many organisms from the 'summer'
phytoplankton, but temperature increases above 20 °C should be avoided.
Suitably illuminated culture flasks standing in a shallow glass tank with tap
water flowing slowly through the tank will counteract the effects of increases in room temperature. Fluorescent strip lights are used as the sole illumination source or as a supplement to daylight. 'Cool white' fluorescent
tubes are popular, although it should be remembered that artificial sources
of illumination can never exactly simulate the spectral composition of
natural light. Controlled environment rooms in which illumination,
temperature and 'day length' can be varied at will obviously offer great
scope for culture studies. Interesting work can still be carried out with a far
less sophisticated set-up, however.

2.8.1 Preparatory cultures

In all work with algal cultures standards of cleanliness with both
glassware and handling should be identical with those required for
bacteriological work.

If a water sample taken from a lake or the sea is enriched by the addition
of nitrate, phosphate and soil extract (Appendix) and then subjected to
light and temperature regimes as outlined above, a dense growth of
phytoplankton organisms will be obtained in a week or so. The relative
numbers of the different organisms bear no relationship to the original
proportions in the sample since the enrichment process will have encouraged the rapid growth of some organisms originally present in very
small quantities. These cultures are the means by which isolates of certain
organisms can be obtained. Organisms obtained by using a townet is
another convenient source of isolates although care is needed to ensure that
no damaged cells are picked out. In the preparation of cultures of brackish-
water organisms it will be necessary to take account of the salinity of the
original water sample (Appendix).

2.8.2 Enrichment cultures

This term has been used to describe cultures in which more specific conditions are imposed in order to encourage the growth of particular
organisms. Factors such as pH control, temperature increase, or the addition of a certain nutrient mixture, are employed. This approach sometimes
requires foreknowledge of the requirements of the organism unless one is

setting out to obtain species typical of certain ecological niches (e.g. growing at a low pH, or with a particular organic carbon source).

2.8.3   Isolation of algae and bacteria-free unialgal cultures

With all successful isolations it must be remembered that a single algal cell brings with it bacterial contaminants. For the reasons given in earlier pages, a *pure* culture of an alga is often required. Removal of the bacterial contaminants calls for a number of measures, some requiring a high degree of technical expertize.

2.8.4   Cell washing, plating, and use of antibiotics

PIPETTE METHOD: This method involves continuation of a dilution technique. The isolated cell (or groups of cells) is transferred through a series of samples of sterile medium, using a sterile pipette (Fig. 2–8a,b) for each transfer. It is an advantage to watch the transfer of motile cells under the microscope so that they will not be lost in the larger volume for the next transfer. Sterile watch-glasses are the most suitable vessels for the washing fluid (Fig. 2–8c). The choice of sterile medium for washing depends on the organisms involved; for phytoplankton about 3 cm³ of mineralized medium is recommended. Six serial transfer 'washings' are regarded as sufficient to remove adherent bacteria.

PHOTOTACTIC BEHAVIOUR   Motile representatives of the phytoplankton can be isolated and cleaned by using their light-sensitivity reactions. If a drop of the material to be cleaned is added to the lighted side of 3 cm³ of 'washing' medium in a watch glass, then after 5–10 min the motile cells will have congregated on the side away from the light. So long as heating of the medium is avoided, window illumination or a shaded bench lamp can be

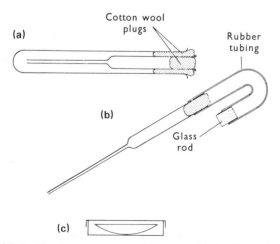

Fig. 2–8   **(a)** Sterile pipette—storage. **(b)** In use with flexible tubing to control movements of liquid. **(c)** Sterile watch glass—storage.

used to light one side. The migrated organisms can be pipetted to the lighted side of another 3 cm^3 of medium and the process repeated. This is an elegant technique. The movement of the organisms through the water means that most of the associated bacteria are left behind. A skilled operator can use this phototactic method to clean individual nanoplankton organisms. 'Picking-out' individual cells of organisms 5–15 μm in diameter would seem impossible but, using a medium power lens in a binocular microscope, the tiny specks moving in the medium can be detected and picked out with a micropipette. Unialgal cultures can be prepared from single-cell isolates of organisms in this size range.

AGAR PLATING   The principles involved here are similar to those used in bacteriological work. A 2% solution of agar in a suitable mineral medium (marine or fresh water) is used, the autoclaved solution being poured into petri-dishes. The sample of algae is taken up in a wire loop and then streaked over the surface of the agar, usually in a zigzag pattern. The petri-dishes are then inverted over or under a light source (Fig. 2–9a). The plating method is used to separate clumps of cells and also to isolate cells in order to obtain clonal growths. Algal cells with associated bacteria will show a cloudy halo. A clean patch of algal cells should then be picked up with a sterile wire loop and the material streaked on a second agar plate, so that a further check can be made. Algal cells cleaned in this way on agar can be transferred to liquid media. Agar slopes in test tubes can be used instead of petri-dishes. A 1.0–1.5% agar solution is then used; this has the advantage that drying out of the agar medium is delayed, but for beginners the plating method with petri-dishes is recommended.

USE OF ANTIBIOTICS   Antibiotics have been used for some twenty years to eliminate bacteria from algal cultures. This method is the easiest to use, but has the disadvantage that undetectable cell injuries can take place. However, with many non-motile phytoplankton organisms and filamentous representatives with mucilage cover, antibiotic treatment is the only workable method (as a general rule, however, washing or plating methods are the first choice). The principle is to subject the algae to high concentrations of a broad spectrum of antibiotics over a short period. This will severely lessen the chances of survival of bacterial cells; most algae are much more tolerant of high antibiotic concentrations than are bacterial cells. DROOP (1967) has described a method of algal culture purification (see summary in Fig. 2–9b). The antibiotic mixtures recommended are:

Antibiotic	*For diatoms*	*For flagellates*
Benzyl penicillin SO$_4$	8000 μg cm^{-3}	8000 μg cm^{-3}
Streptomycin SO$_4$	1600 μg cm^{-3}	2000 μg cm^{-3}
Chloramphenicol	200 μg cm^{-3}	8 μg cm^{-3}

Six tubes containing 6 cm^3 algal culture are used, each with a dense and vigorous algal growth. The freshly prepared antibiotic mixture (12–15 cm^3) is filter-sterilized and 6 cm^3 used. The antibiotic tube is then mixed

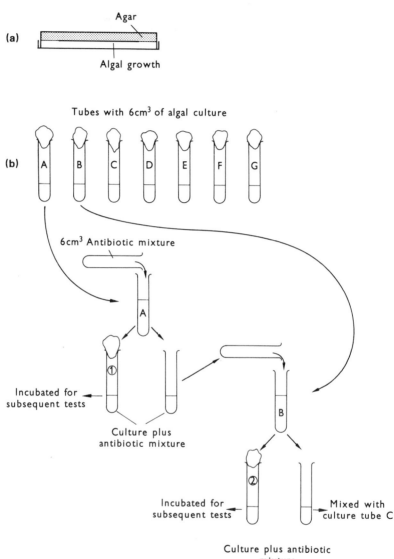

Fig. 2–9 **(a)** Algal 'streak' on nutrient agar; incubated with light source above or below dish. **(b)** Diagrammatic summary of method of purifying algal cultures with antibiotics.

with one of the cultures over a flame with the result that two tubes are obtained, each with half the original antibiotic concentration. One of these tubes is marked '1' and set aside, and the other mixed in the same way with a second culture tube. Of the two tubes so obtained one is marked '2' and the other mixed with the third algal culture. After using all six culture tubes, a series of 1–6 is obtained in which the antibiotic mixture is decreased by a factor of two in successive tubes. One drop of sterility test medium is then added to encourage growth of bacteria, since penicillin acts on the dividing cell. After 24 hours subcultures are made into sterile antibiotic-free media, with the subcultures marked 1–6 to indicate the type of pretreatment. A variety of media will probably be necessary at this stage, since it cannot be assumed that the medium successfully used with the alga before antibiotic treatment will again be suitable for growth under bacteria-free (axenic) conditions. Sterility tests (see below) should be made on the subcultures after three weeks. Whilst results may not be generally applicable, it could be expected that tubes 5 and 6 (and possibly 4) would still contain bacteria, and tubes 3 to 6 (and possibly 2) would contain living algae. Tubes 3 and 4 would be the best choice for further subcultures if bacteria-free cultures are to be obtained. One caution: with antibiotic treatments of algae it is seldom difficult to remove the bacterial contaminants. It is invariably much more difficult to get the algal cells to grow after treatment. Hence the need to try a number of media for axenic growth.

STERILITY TESTS No matter what the method of elimination of bacteria, it can never be assumed that the treatment alone is a guarantee of success. Regular tests to check the continued sterility of the subcultures are essential; the standard procedure is to add an organic substance which in normal circumstances would encourage proliferation of the bacterial cells. Minimal bacterial contamination of a subculture will then become evident in the cloudy appearance of the cultures after a short period of incubation (24 h). The organic mixture most frequently used is a bacto-peptone preparation (0.1%). Hence in all serial subcultures made after any of the cleaning methods some of the subcultures must be set aside for sterility checks.

3 Buoyancy of Phytoplankton

The light dependency of photosynthesis and the attenuation of incident light in its passage through water indicate limitations on phytoplankton metabolism should organisms sink into regions inadequately illuminated. A stationary cell will be at a disadvantage because it will absorb the nutrients in its contact layer; for the cell to remain bathed in a continuing supply of dissolved nutrients it must keep moving, and the natural tendency is for phytoplankton cells to sink. If, however, there is daily residence in the lighted zone long enough overall to allow carbon fixation sufficient to satisfy the metabolic needs of the cell, there will be no stress effects—even though the periods of illumination may not be continuous. Periodicity of access to the lighted zone will require some slowing down or reversal of the downward drift. With flagellum-bearing organisms there will be appreciable energy expenditure if flagellum action is to be the sole means of remaining in suspension. Too rapid a fall into the aphotic zone may prove lethal unless there is time for formation of resting spores or other resistant phases. The sedimentary bottom deposits of diatom shells, coccolithophorid scales and silicoflagellate skeletons illustrate the steady drain of surface organisms into the depths.

3.1 Phytoplankton suspension

Phytoplankton suspension is the result of cells being buoyed up by a force equal to the weight of water displaced minus the weight of the cell (SMAYDA 1970) e.g.

$$F = gkd^3 (\rho^1 - \rho)$$

Where F = resultant force of gravity
$\qquad g$ = acceleration due to gravity
$\qquad kd^3$ = cell (or chain) volume (d = a linear measurement)
$\qquad \rho^1$ = density of organism
$\qquad \rho$ = density of medium
$(\rho^1 - \rho)$ = excess density

When the excess density is zero, or, if positive, there is a compensatory external force, the organism remains in suspension. The density of cytoplasm of marine organisms has been calculated as lying between 1.03 and 1.10 g cm^{-3}. Seawater densities range from 1.021 to 1.028 g cm^{-3}. Phytoplankton organisms bear additional ballast (silica, calcium carbonate, cellulose). Hence in both sea and freshwater plant-cell densities are appreciably more than that of the medium. The paradox is that most phytoplankton organisms are too heavy ($\rho^1 > \rho$), and there must be some compensation for this overweight if they are to remain in suspension. Bloom-forming freshwater blue-green algae and the marine blue-green alga *Trichodesmium* show positive buoyancies ($\rho^1 < \rho$), attributed to gas-vacuole formation (p. 46).

3.2 Morphology of phytoplankton cells and buoyancy

It is often implied that selection processes in the evolution of phytoplankton organisms have resulted in morphological forms which in themselves constitute flotation adaptations (see Fig. 3–1). For diatoms these are bladders, plates or ribbons, needle or hair-like cells, and cells with hair-like projections. Bladder-like cells have thin silica walls and large vacuoles with cell-sap of low density. Needle-like cells sink slowly if their long axis is horizontal to gravitational pull, but rapidly if vertical. Ribbon-like colonies are broad and flat, and show some twisting in suspension.

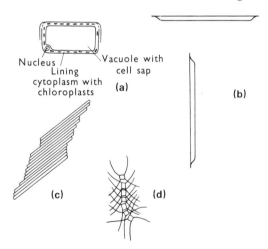

Fig. 3–1 **(a)** Bladder-like cell with large vacuole (*Coscinodiscus*). **(b)** Needle-like cell (*Rhizosolenia*)—floating position (upper) and sinking position. **(c)** Raft-like cell mass (*Bacillaria*). **(d)** Associated group of cells with spiny outgrowths (*Chaetoceros*).

Hair-like cell outgrowths are also said to offer resistance to sinking. The spiny and wing-like outgrowths of some dinoflagellate cell walls have been described as flotation devices which supplement flagellum activity. Stokes Law for sinking bodies states that the slowest rates of fall will be obtained with those bodies in which the surface-area : volume ratios are greatest—conditions which entail a greater frictional drag with the water. Theoretical interpretations, with particular reference to cell forms in phytoplankton, have shown that the relationship between shape and sinking rate is directly linked with size (MUNK and RILEY, 1952). For organisms of 5 μm diameter the order of decreasing sinking rates is plate > cylinder > sphere; with organisms of 50 μm diameter the relationship is cylinder = plate > sphere, and with diameters of 500 μm, cylinder > sphere > plate. In general, however, increase in cell size results in an increased sinking rate irrespective of cell shape. Rate of sinking also increases with ageing of the cells, presumably because of physiological changes which increase the density of the cell sap or cytoplasm. Cell aggregations or colony formation by planktonic algae are also considered to be flotation adaptations. Spherical colonies of small cells of low density embedded in a mass of mucilage (e.g. *Phaeocystis*) will clearly approach conditions of positive byoyancy. This is exemplified by the seasonal persistence of *Phaeocystis* 'patches' in the surface waters of north temperate seas. Colony formation by planktonic diatoms, although sometimes described as a flotation adaptation, invariably results in increased sinking rates. An exception here is the common marine diatom *Skeletonema costatum*. Increased colony sizes have apparently resulted in decreased rates of sinking. Colony form is quite distinctive in this species, consisting of a linear series of small cells linked by minute silica rods. These microscopic rods may induce microturbulence in their immediate vincities, so counteracting the tendency to sink (SMAYDA, 1970). The density difference between mucilage and the organism must be at least twice the difference between the densities of mucilage and medium if such a covering is to be an effective buoyancy mechanism. It is unlikely that the mucilage covering of diatom colonies will be large enough to be effective. Colony formation in diatoms may be a result of a rapid division rate leading to cell aggregations less palatable to zooplankton. Diatom colonies show seasonal and regional variations in form but there appears to be little practical evidence that these are flotation adaptations.

A variety of superficial outgrowths and protuberances are borne by phytoplankton cells. These give a high surface-area to volume ratio and increase the frictional drag between the organism and the water, but if they are silicified or calcified the excess density would counteract any buoyancy advantages gained through increased frictional drag. *Scenedesmus* is a green alga with aggregations of up to four flattened cells with spiny outgrowths. Whilst the plants are common in shallow freshwater pools, they are 'practically never observed in the plankton of lakes as large and deep as Windermere' (LUND, 1959). The elaborately incised disc-like cells of *Micrasterias* spp. (Fig. 1–6) would similarly appear well adapted to planktonic life on the basis of a high surface-area to volume ratio and yet these are also rare components of lake phytoplankton. Spiny desmids (e.g. *Staurastrum* spp.) often occur in lakes; if these spiny organisms are examined when immersed in indian-ink suspension the outermost mucilaginous cover often gives them an overall spherical appearance. Such a mucilage envelope will help to reduce the cells specific gravity. Spiny outgrowths on desmids could also reduce the chances of the cells being eaten by animals. Surface outgrowths and protuberances on phytoplankton cells can be interpreted as anti-predatory devices, as a means of enlarging the absorptive area for uptake, and as cell-orientating devices in some genera (e.g. *Rhizosolenia* spp., Fig. 4–4). Whether they serve also to improve buoyancy by frictional effects with the water will depend on the overall ballast borne by the cell. Where unicellular organisms bear thread-like protoplasmic outgrowths, (e.g. radiolarians or heliozoa), some flotation properties can be envisaged.

It is well known that loss of flagella or cessation of their beat results in immediate sinking of algal cells. Organisms classified as ultra- and nanoplankton are of a size range in which flagellar motility as an aid to flotation can be regarded as superfluous. There are, however, numerous flagellated nanoplankton organisms which bear coccoliths and siliceous scales, and mineralized structures of this nature will induce excess density conditions. Thus the coccolith-bearing organism *Coccolithus huxleyi* (which lacks flagella) has been shown to sink at a rate five times that of a coccolith-free variant of similar cell size. Recent work with the electron microscope has shown that scales of a polysaccharide nature are borne by green and brown-coloured nanoplankton organisms. These scales often show a remarkable degree of intricacy of form and ornamentation. Nothing is known about the effects of these organic scale coverings on cell density, nor of their functional significance. Organic scales which are spiny, or in any way raised up from the cell surface, may limit chances of predation. In some green-coloured flagellates organic scales are also borne on the flagella, but the functional significance of this is difficult to explain.

Dinoflagellates have been the main source of studies on flagella and sinking rates. In addition to the specialized arrangement of the flagella (p. 10), there are horn-like and wing-like outgrowths of the cell walls in some species (Fig. 1–5). The more elaborate the surface expansions the more swimming speed is reduced. If these cell wall extensions serve to increase form resistance (and possibly have anti-predation functions as well) and assist in maintaining the cells in suspension, then flagellar action as an aid to suspension would be superfluous.

3.3   Physiological regulation of cell density

An intracellular control of cell density would seem an effective means of regulating the sinking process. Included here are reserve substances, gas vacuoles, and ionic composition of the cell sap.

3.3.1   *Fat reserves*

Fat accumulations in phytoplankton cells are regarded as means of counter-balancing the excess density due to mineralized cell walls. In diatoms appreciable fat deposition regularly occurs in species with heavily silicified cell walls. Appreciable fat deposits in cells are not necessarily indications of buoyancy adaptations, however. It is known that accumulations of fat develop in cells under certain stress conditions (e.g. interruptions in nitrogen metabolism; nitrate deficiency in the water; high light intensity stresses, and ageing of the cells). Fat accumulations also accompany breakdown of cell organization. It would seem odd if diatom cells were most buoyant at times when they were least viable. Ageing diatom cells in fact sink more rapidly. However, fat laden cells of the freshwater diatom *Nitzschia palea* were found to sink at an appreciably slower rate than normal cells lacking these extensive fat deposits. Whether these fat-laden cells were senescent is unknown. Colonies of the freshwater organism *Botryococcus braunii* show marked flotation properties, and this is attributed to its high lipid content (30–40% of dry weight). This organism also forms aliphatic hydrocarbons which will increase buoyancy.

3.3.2   Gas vacuoles

Surface 'blooms' of blue green algae are a well known feature of lakes and ponds in periods of calm weather (p. 79). These blooms are commonly of filamentous and colonial species whose cells contain gas vacuoles. Gas vacuoles are gas-filled spaces within the gel-like cell protoplasm, bounded by cytoplasm which confers some rigidity to the vacuole shape. In each vacuole there are a number of minute gas cylinders or vesicles which can be isolated from the cells without loss of integrity. Each cylinder is in fact kept distended by minute hoop-like proteinaceous structures. The rigidity of the membrane allows the vacuole to withstand pressure changes caused by movements of the cells in a water column. The cell processes which lead to vacuole formation are unknown, but the gas mixture in them is known to resemble that in the water (mainly nitrogen mixed with argon and occasionally oxygen), and would seem not to be specific products of metabolic activity. The buoyancy properties conferred on the filaments by gas vacuoles can be readily demonstrated if the vacuole membranes are ruptured by a sudden high pressure shock. Filaments so treated sink rapidly to the bottom of the container. Gas vacuoles also occur in blue-green algae which are not planktonic, and in certain of these species vacuole formation is induced by exposure to bright light. Other functions suggested for gas vacuoles have included light-shielding properties, metabolic by-products and residues, and bubbles of gas produced during periods of anaerobiosis. But the gas mixture present is unlikely to be produced during anaerobiosis. Whilst their contributions to the buoyancy of planktonic blue-green algae is clear-cut, this may be incidental. Under turbulent conditions buoyant cells are at a considerable advantage. With long periods of calm weather, however, it can hardly be advantageous for vast numbers of organisms to be crowded at the water surface with all the competitive factors that arise. Further, mild wind action will often result in the surface 'bloom' of algae being deposited on the shoreline. Not all blue-green algae with gas vacuoles show such positive buoyancy, however. Examples are known where organisms remain for long periods in the vicinity of the metalimnion of lakes, maintaining this position by metabolic control. Gas vacuole formation is largely a feature of freshwater blue green algae, and is rare in marine representatives. The bloom-forming *Trichodesmium erythraeum* is the best known example.

3.3.3   Control of ionic composition of cell sap

In diatoms of the 'bladder' type (p. 43) a great part of the cell volume is vacuole and cell sap (vacuole volume has been estimated as 50% of the total volume of cells of *Lauderia annulata*). GROSS and ZEUTHEN (1948), in a study of the buoyancy of the marine diatom *Ditylum brightwellii* stated '... under suitable physical conditions marine plankton diatoms do not sink because their specific gravity is equal to that of sea water'. Since so much of the weight of the diatom cell is in the wall the vacuolar sap specific gravity would have to be appreciably reduced to compensate for this. *Ditylum* vacuole sap specific gravity was estimated as 0.0025 less than that of sea water when cells were growing and remaining in suspension. Actively growing *Ditylum* cells were said to have a neutral buoyancy. *Ditylum* resting spores were observed to sink rapidly, and a similar result was obtained if any check occurred on cell growth. Resting-spore formation is accompanied by contraction of cell contents and loss of cell sap. If growth of cells is stopped then it seems that the ionic composition of the vacuole sap comes into equilibrium with sea water and a negative buoyancy of the cell results. Gross and Zeuthen also showed that with two isotonic sea waters with one deficient in divalent cations and anions (Ca^{2+}, Mg^{2+}, SO_4^{2-}) had similar differences in specific gravity to those between diatom cell sap and sea water of salinity 32.5 parts of salt per thousand. On the basis both of experimental studies on sinking rates and cell densities, and on derived information regarding the cell sap composition of other marine algae, Gross and Zeuthen suggested that the neutral buoyancy of *Ditylum* was in part due to the selective absorption of monovalent ions (e.g. Na^+ and K^+) and the maintenance of low concentrations of divalent ions, both processes involving energy expenditure on the part of the cell.

The holozoic dinoflagellate *Noctiluca miliaris* accumulates in vast numbers at the sea surface during calm weather and apparently regulates its density with accompanying volume changes. Cell-sap analyses show very low concentrations of divalent ions, a high concentration of Na^+ relative to K^+, and quite high concentrations of ammonium (NH_4^+) ions. The buoyancy of this species would seem to be due to a selective accumulation of light monovalent ions and low concentrations of divalent ions (particularly SO_4^{2-}). Ammonium ions are unlikely to accumulate in autotrophic plant cells which are nutritionally dependent on exogenous sources of dissolved combined nitrogen.

Gross and Zeuthen's explanation of diatom buoyancy has no relevance to freshwater diatoms. The dissolved salt content of Lake Windermere is very low (e.g. 33 mg dm^{-3} total ions; that of sea water is a thousandfold greater). Colonies of *Asterionella formosa* never have the same density as lake water and invariably sink in the absence of any turbulence (LUND, 1959). Gross and Zeuthen provided no analysis of *Ditylum* cell sap in support of their theory, but with the technical difficulties involved this is hardly surprising. However, the 'giant' marine diatom *Ethmodiscus rex,* with sap volumes ranging from 0.45–1.45 mm³, has been analysed. In healthy cells there was a tenfold concentration of Na^+ over K^+ and a complete absence of NH_4^+. Ca^{2+} ions were present in very small quantities, and Mg^{2+} ions were not detected. SO_4^{2-} ions were present in similar quantities to K^+ ions in living cells, but SO_4^{2-} ion concentrations increased tenfold in damaged and dead cells. These results support the hypothesis that exclusion of divalent ions is a feature of the cell sap of planktonic diatoms. Of equal importance, however, is whether this lowered specific gravity of the cell sap is sufficient to counterbalance the silica 'ballast' of the cell.

There would appear to be insurmountable problems in resolving this. To reconcile the variable shapes of diatom cells with absolute measurements seems an impossible task. Such an attempt has been made (SMAYDA 1970). The silica 'ballast' of a *Ditylum* cell has been estimated as 0.1% of the total cell density but only 0.05% of the total cell volume. Each 1 μm³ of silica wall requires 2.5 μm³ of cell sap of density 1.02 to ensure buoyancy. Hence it is estimated that the ionic composition of the cell sap becomes a significant feature when there is a surface-area to vacuole-volume ratio of 0.45 and below. High surface-area to volume ratios are obtained with small cells, so that selective ion accumulation and exclusion would seem inapplicable to small diatoms. Selective ion exclusion has also been found to occur in a marine green alga which is not planktonic (*Valonia*), and in the buoyant spherical organism *Halicystis*. In each, low concentrations of divalent ions occur, but the buoyant *Halicystis* has also a much higher concentration of Na^+ than K^+. Gross and Zeuthen's hypothesis appears to be a plausible explanation of flotation for some marine diatoms, although not generally applicable.

3.4   Physical factors due to surrounding medium

The two principal factors are water movement and viscosity. The viscosity of water decreases with increasing temperature and in tropical waters this would clearly make plankton suspension more difficult in the warmer months. Organic substances liberated by phytoplankton would also affect water viscosity, particularly during periods of 'bloom'. In addition these substances would influence the water turbulence ('oil on troubled waters') and so indirectly affect suspension. This aspect of phytoplankton ecology has been little explored.

Water turbulence has a considerable influence on phytoplankton. No matter how calm a water mass appears, it is likely that small-scale (even microscale) turbulence is taking place. Turbulence arises from a number of sources. Vertical eddy diffusion currents are reflected upwards from water flowing over uneven areas of sea or lake bed. Tilting of the discontinuity layer in lakes through surface wind action and the see-sawing action of the seiche will induce turbulence above and below the metalimnion. Wind-induced 'convection cells', localized regions of downwelling and upwelling, are known in ocean waters. Similar convection cells have been reported from lakes. Wind of varying force and direction is a persistent feature; even when minimal there is still a significant effect on water movement and surface turbulence. Thus there is much support for the contention that water turbulence, allied with certain of the flotation mechanisms already described, may largely account for phytoplankton suspension in the sea and lakes.

3.5   Field and laboratory observations

There is evidence that sinking rates of marine and fresh water diatoms are much the same in nature. With flat calm and minimal water disturbance the tendency is for most organisms to sink. Exceptions are seen in gas-vacuole-containing blue-green algae and organisms such as *Botrycoccus* (p. 46). Healthy cells of the small freshwater diatom *Cyclotella melosiroides* show a very slow sinking rate, whereas *Melosira italica* subsp. *subartica* and *Asterionella formosa* sink rapidly in calm water, *Melosira* 3–5 times faster than *Asterionella*. *Melosira* almost disappears from the plankton with the onset of stratification. Its period of maximum abundance is between autumn and late spring. *Asterionella* similarly shows maximum development under turbulent conditions. With summer thermal stratification in lakes depletion of diatom populations will occur if the rate of loss from the epilimnion by sinking is greater than can be contained by multiplication and growth of the organisms.

Cells of *Asterionella* and *Melosira,* and filaments of the blue-green alga *Oscillatoria agardhii* var. *isothrix,* all tend to assume a vertical position in still waters—a position which presents the least surface area to the vertical axis of the water column and will lead to maximum rates of fall. Entirely still conditions rarely exist in nature but shapes of algal cells may be significant in slowing down the time available for re-orientation. Cells of *Rhizosolenia* lacking terminal spines (Fig. 4–4) sink faster than the spine-bearing forms, and those with one spine faster than those with two. Spiny outgrowths on *Chaetoceros* cells induce slow rotational movements of the whole chain when falling through the water. This slow rotation will help in suspension, and also equalize illumination and prevent chloroplast damage. STEELE and YENTSCH (1960) observed that a chlorophyll maximum was obtainable at the bottom of the euphotic zone in the sea. This was attributed to a slowing down of the sinking rate, the cells increasing their buoyancy in the dark nutrient-rich waters. The sinking rates of aged *Skeletonema* cells and starved *Rhizosolenia setigera* were appreciably slowed down after transference to nutrient-rich media. Buoyancy would appear to be closely linked with the general physiological condition of the cells.

3.6   Sinking, rising and rotating

Diatom cell movements increase the opportunities for uptake of nutrients by processes of forced convection, which ensure the continuing renewal of the cell contact layer. The rate of nutrient transference will depend on cell morphology and the properties of the surrounding water. Any movement—including sinking—will be advantageous. Upward movement (as with blue-green algae containing gas vacuoles) will enable forced convection to take place. Culture experiments have shown that increased sinking rates accompany nutrient depletion. If cells pass out of the photic zone for short periods they are protected from stress effects due to continued photosynthesis in the absence of sufficient nutrients.

4 Successions and Associations

Seasonal changes in the terrestrial vegetation of temperate latitudes are well known. Seasonal changes in aquatic habitats are less obvious. In the marginal areas of lakes the rooted vegetation shows seasonal changes similar to those of terrestrial plants. To the casual observer the narrow coastal fringe of seaweeds changes little throughout the year. The seasonal successions of phytoplankton often appear as reflected colours. The surface colours of some lakes are white or metallic grey in spring changing to green, thence to orange and gold before darkening again in late autumn and early winter (HUTCHINSON, 1967). In the sea, noticeably in coastal waters, a variety of 'fisherman's signs' result from plankton changes. At night in the summer months phosphorescence of surface waters is due to the abundance of luminescent dinoflagellates. Water blooms of blue-green algae which appear on the surface of productive lakes during summer and autumn are more dramatic manifestations of complicated physico-chemical changes in the water mass.

4.1 Spring outbursts

Different annual patterns of phytoplankton abundance depending on latitude are to be seen in the sea (Fig. 4–1a–e). Increases in phytoplankton populations in the spring are characteristic features of north temperate seas and many lakes. This phenomenon is described either as the 'spring outburst' or 'vernal blooming'. Interpretations of the temporal aspects of phytoplankton biology are very much dependent on the frequency and regularity of sampling programmes. Much of the information now available has been obtained in weekly collections. Some reports are based on observations with longer intervals. The irregularity of sampling is inevitably due to the physical difficulties of getting collections more frequently. Weekly sampling programmes allow inferences to be made regarding the periodicity of the larger organisms; the waxing and waning of some nanoplankton organisms in this interval may be missed. Microbial phytoflagellates can show marked population increases in a few hours. Even daily sampling may give an inadequate picture for inshore coastal regions where tidal influences have to be taken into account. Here diurnal population changes have been descibed from short-term observation. An ideal sampling programme for large lakes or the sea that involves all biological, chemical and physical parameters may well prove unworkable in terms of manpower and cost; problems of data handling would also arise with such intensive studies. Much relevant information has been obtained despite the acknowledged imperfections of sampling programmes. The broad picture of events has been known for many years, and hypotheses advanced about seasonal successions. It is a sobering thought, however, that the controlling factors of many obvious seasonal changes are still inadequately understood.

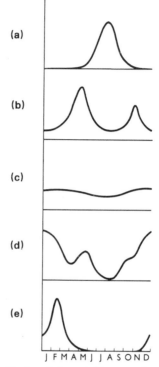

Fig. 4–1   Seasonal amplitudes of phytoplankton production. **(a)** Arctic seas. **(b)** North temperate seas. **(c)** Tropical seas. **(d)** Antarctic seas—northern region. **(e)** Antarctic seas—southern region.

Spring outbursts are sufficiently regular in some localities for their occurrence to be timed and even predicted. Predictions, however, often prove unreliable—to be expected with a sequence of events governed by the vagaries of climate. Nutrient levels are high during winter in lakes and north temperate seas. Whilst phytoplankton populations are sparse laboratory incubation of winter lake-water samples at higher temperatures and light intensities yields dense algal growths. In winter the division rates of the diatom *Asterionella formosa* in Lake Windermere may fail to keep pace with loss of cells—largely due to the outflow of water from the lake. This is thought to be the most significant factor in regulating the winter population of this species. Temperature changes in the water are small at the time of onset of spring outbursts. A spring growth will take place under ice if illumination is sufficient. There is a time lapse between maximum insolation and maximum water temperatures. The insufficiency of winter sunlight and shorter days in middle and higher latitudes are the main factors restraining algal growth. Cell multiplication will only occur when production of organic matter by photosynthesis exceeds respiratory breakdown. If the top layer of mixed water is deep and there is even distribution of organisms from the surface down, net production of organic matter will only occur above the compensation depth. Organisms will be carried out of the photic zone too rapidly for cells to multiply. Below this there will be net loss due to respiratory breakdown. Overall population cannot increase if loss exceeds net production. There is a critical depth for the mixed layer where production will be at a standstill. If the depth of the mixed layer is less than this and there is adequate illumination, the onset of the spring outburst is ensured. This critical depth has been estimated as 5–10 times the compensation depth. There is evidence that a period of calm weather (2–3 weeks) precedes the onset of the spring outburst in some localities (e.g. the Firth of Clyde). The significance of light as principal regional factor influencing the spring growth is now well documented. Winter restraint on population increase is largely a result of the shallowness of the photic zone relative to the depth of mixed water; lowered temperatures may be an accompanying factor.

Once under way, the spring outburst of phytoplankton can be likened to the growth of a microbial population in culture. The rate of increase in nature follows roughly an exponential course over several weeks. This is an overall population increase incorporating successive crops of individuals which wax and wane. Division rates in nature vary—estimated as ranging from 2–10 times per month depending on the species in some lakes. This process is distinct from the growth and decay of a single generation of annual plants on land. Annual changes in lake phytoplankton have been compared with changes in forest cover during a period of postglacial climatic change. We lack knowledge of the precise factors governing the waxing and waning of successive crops of individuals, including the origins of species which produce the successive crops.

As shown in Fig. 4–1b, the spring outburst is short and a decline in population soon sets in. In north temperate seas this is mainly attributed to the grazing of herbivorous zooplankton (p. 72). The peak of zooplankton production lags behind that of the algae by a few weeks. Eventually the grazing rate will exceed that of algal cell division. The relatively shallow mixed layer confines the algae to a narrow zone with increased illumination but steadily declining mineral nutrients. The zooplankton appear often to feed at rates bordering on 'luxury consumption' (or 'superfluous feeding, p. 73), voiding faecal pellets containing partially broken and sometimes undigested diatom cells. These drift down into the deeper water and decompose. Later, when the available phytoplankton is markedly reduced the remaining faecal pellets in the photic zone may serve as food. With thermal stratification in summer the isolation of surface waters is complete. Overall algal production remains small except for occasional 'pulses'. Thermal stratification effectively locks in deeper water the nutrients liberated by bacterial action on decomposing organic matter and zooplankton faecal pellets. The early summer decline in phytoplankton in certain lakes is not so directly attributable to zooplankton grazing because the larger diatoms and dinoflagellates are less subject to predation (p. 77). That the peak of spring growth coincides with the depletion of silicate in Lake Windermere may not fully explain the consequent population decline; exhaustion of other nutrients (organic or trace elements), overcrowding (p. 60), and conditioning by preceding organisms (p. 65) may play some part. In some lakes the onset of stratification is accompanied by a mass death of diatoms in an epilimnion now depleted of nutrients. Some viable cells remain, however.

4.2   Species succession in spring outbursts

A cumulative graphical summary of a spring outburst tells us nothing about its most interesting features. Over the period there is a steady succession of dominant and codominant organisms—a sequence of waxing and waning, often involving numerous species. The most fascinating aspect of such a study is to watch the day-to-day changes in species observable in large water masses. In small ponds fluctuations in numbers and species of plankton can be observed over shorter intervals. It would be wrong to conclude that plankton cycles observed over 1–3 years are typical of more than that period. It is also unwise to assume that the cycle of events described for one lake is typical of all lakes of similar size—even in the same area. Data for Lake Windermere apply only to that lake (LUND, 1964).

4.2.1   The sea

MARGALEF (1958) has described phytoplankton succession as a sequential process starting with small-celled diatoms capable of a high photosynthetic rate and rapid cell division and hence requiring high nutrient levels. These are followed by slower-growing diatoms of medium size, and then by motile forms (dinoflagellates) and other organisms with more complex nutritional requirements (e.g. blue-green algae). The origins of species which appear in the first stages of the spring outburst are also important. Whether they arise from 'stocks' already present in small numbers, or from a perennating phase, or whether 'seeded' from other areas, will have a clear bearing on the succession. In some north temperate coastal localities certain diatoms are common in the first stages in the spring outburst. These include *Thalassiosira* spp., *Chaetoceros* spp. and *Skeletonema costatum* (Fig. 4–2a–c). Species of *Coscinodiscus* and *Biddulphia* are also prominent. In some localities *Skeletonema* often dominates the spring outburst (e.g. in the Firth of Clyde and Southampton Water). At Whitstable in the Thames Estuary *Biddulphia aurita* (Fig.

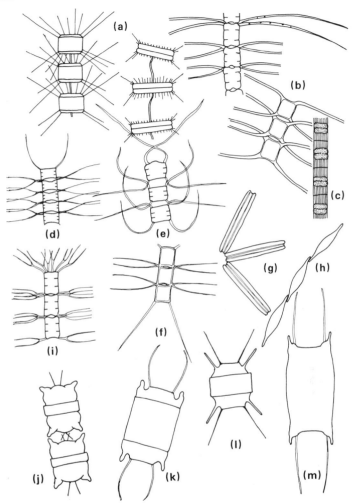

Fig. 4–2 **(a)** *Thalassiosira nordenskioldii* (left) and *Thalassiosira decipiens* (right). **(b)** *Chaetoceros densum* (upper) and *Chaetoceros danicum* (lower) **(c)** *Skeletonema costatum.* **(d)** *Chaetoceros decipiens.* **(e)** *Chaetoceros laciniosum.* **(f)** *Chaetoceros affine.* **(g)** *Thalassiothrix nitzschioides.* **(h)** *Nitzschia seriata.* **(i)** *Bacteriastrum delicatulum.* **(j)** *Biddulphia aurita.* **(k)** *Biddulphia regia.* **(l)** *Biddulphia mobiliensis.* **(m)** *Biddulphia sinensis.*

4–2j), together with smaller quantities of *Skeletonema*, has been described as the dominant organism in the early spring flowering. In oceanic waters the dominant diatom species often differ from those near the coast. Two years' observations from weather ships I and J in the North Atlantic showed *Chaetoceros decipiens C. laciniosum, C. affine, Thalassiothrix nitzschioides, Nitzschia seriata* and *Bacteriastrum delicatulum* (Fig. 4–2d–i) to be the dominant species. All species so far mentioned probably represent the small-celled, quick growing organisms of the first phase of a spring outburst.

From data accumulated over many years, attempts have been made to correlate seasonal occurrences and certain 'indicator' species. Thus for the north-west European seas the following diatoms are listed as being confined mainly to the spring flowering period: *Achnanthes taeniata, Chaetoceros atlanticum, Chaetoceros cinctum, Chaetoceros diadema, Chaetoceros furcellatum, Chaetoceros holsaticum, Chaetoceros teres, Chaetoceros wighami, Coscinosira polychorda, Lauderia glacialis, Leptocylindrus danicus, Melosira hyperborea, Navicula vanhoffenii, Nitzschia frigida, Rhizosolenia hebetata, Thalassiosira gravida, T. nordenskioldii.*

A ruling that these species are to be found only in the spring assumes a constancy of climatic factors unlikely to occur.

In the Clyde Sea area successive crops of diatoms in the spring succession have been observed to sink to the deeper layers as they wane (Marshall, in HARVEY, 1963).

4.2.2   Lakes and ponds

Spring outbursts have been described for many lakes and ponds. In Lake Windermere the diatoms *Asterionella formosa* (Fig. 4–3a), *Tabellaria fenestrata* var. *asterionelloides* (Fig. 4–3b) and *Fragilaria crotonensis* (Fig. 4–3c) tend to dominate the spring flora, and usually commence growth at about the same time, although the main growth of *Fragilaria* lags behind the other two (April, May, June). The three species reach their maximum when available silicate is exhausted. The diatom *Melosira italica* (Fig. 4–3d,e) shows a growth period not synchronous with the spring outburst. Its maximum is reached in early March before the available silicate is used up. Desmids show an increase at about the same time although their main period of abundance is in summer. There thus seems to be no initial competition between diatoms and desmids. In Lake Mendota (Wisconsin) the diatom *Stephanodiscus astraea* (Fig. 4–3f) developed enormous populations in March and April during 1915–1917. *Asterionella, Tabellaria* and *Fragilaria* showed similar times of appearance in spring, but not in the numbers shown by *Stephanodiscus*. *Stephanodiscus* was limited to the spring outburst; the other three diatoms were present most of the year. Long-term studies on Lake Erie have shown changes in dominance of the diatom flora at the spring outburst. Over eighteen years (1931–1949) *Asterionella formosa* dominated the spring outbursts except when it was co-dominant with *Melosira* (1932), and when it was replaced by *Melosira* and *Synedra* (1937). After 1949 *Asterionella* was never again dominant in spring. These changes in the spring flora are ascribed to eutrophication (p. 97). The sequences so far described are more typical of eutrophic lakes. In oligotrophic lakes the diatoms

Cyclotella (Fig. 4–3g) and *Tabellaria flocculosa* (Fig. 4–3b) will often dominate the spring outburst. The numbers are sparse compared with eutrophic lakes. Even when numbers are relatively small it is possible to detect an increase as the spring progresses. In acid moorland lakes it is also possible to see an improvement in the condition of *Tabellaria,* as shown by darkening of chloroplast colour. The sizeable desmid flora is a feature of some oligotrophic lakes (p. 71).

In ponds microbial green algae are often conspicuous components of the spring outburst in addition to diatoms. In Abbot's Pool, Somerset, major peaks of microbial green algae were recorded during the spring and

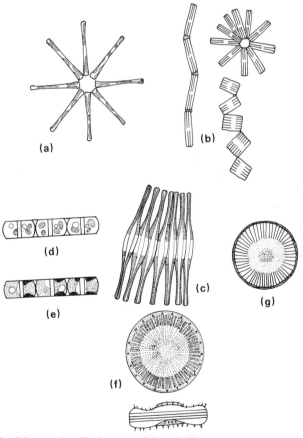

Fig. 4–3 (a) *Asterionella formosa.* **(b)** *Tabellaria fenestrata* (upper left and right) and *Tabellaria flocculosa* (lower). **(c)** *Fragilaria crotonensis.* **(d)** *Melosira italica*—resting stage. **(e)** *Melosira italica*—after exposure to light. **(f)** *Staphanodiscus astraea*—valve view (upper) and girdle view. **(g)** *Cyclotella* sp.—valve view. (**d–e** Modified after LUND, 1954.)

summer of 1966 and 1967 (Happey, in ROUND, 1971). These included *Chlamydomonas* and *Chlorella*. The colonial green alga *Pandorina morum* formed an abundant growth in midsummer.

Each of the freshwater successions described represent the small, rapidly-dividing algae of the first phase of Margalef's sequence. Termination of the outburst coincides with the formation of the thermocline, as well as silicate depletion. This event has been described as a 'shock' period, one in which the lake becomes a shallow area as regards phytoplankton growth and survival, due to stratification (ROUND, 1971). The algae now circulate in the epilimnion, in a limited volume of water and subject to higher temperatures, increased light intensity and a restricted supply of nutrients.

Some of the broader environmental features conducive to a spring outburst were described earlier. These included a high level of nutrients, increasing day lengths and light intensities, and the depth of the mixed layers. The importance of temperature has also been stressed by some authorities. The species differences between winter and spring populations have been put down to algae favouring low temperatures and weak light giving place to those growing best in cold waters but in stronger light. The occurrence of some species has been linked with surface temperature conditions (e.g. Table 5 for the Gulf of Maine).

Table 5 Sea surface temperatures and occurrence of diatoms in the Gulf of Maine. (data from SVERDRUP, JOHNSON and FLEMING, 1946)

month/diatoms	Surface temperature (°C)
April	3
Thalassiosira nordenskioldii	
Porosira glacialis	
Chaetoceros diadema	
May	6
Chaetoceros debilis	
June	9
Chaetoceros compressus	
August	12
Chaetoceros constrictus	
Skeletonema costatum	

It is not easy to accept that temperature changes alone will be the governing factors in a species succession. Here laboratory experiments are not altogether useful. Often the experimental tolerance ranges are appreciably greater than those to which the organism is subjected in nature. Data from field results is sometimes confusing. Thus the linking of *Skeletonema costatum* (Table 5) with surface temperatures of 12 °C (summer) in the

Gulf of Maine has to be balanced against this same species dominating the spring outburst over most years in the Clyde Sea area with sea temperatures of 6–7 °C. Diatoms which are prominent in the spring outburst often appear again in the autumn when temperatures are higher. One interesting observation is the occurrence of two ecotypes of *Asterionella formosa* differing in size and temperature tolerances. These two ecotypes have been found at different depths. The optimal temperature range of the larger ecotype found in the epilimnion is 5.1–8.0 °C, whilst the other (in the hypolimnion) develops sizeable populations in temperatures between 6.7–14.2 °C. It is possible that different 'strains' of a species may participate in the spring and autumn growths. On the other hand it is equally possible that some phytoplankton are more tolerant of wider temperature and light fluctuations than others. Organisms with the wide-ranging tolerances are the opportunists, able to take speedy advantage of favourable conditions and to develop large populations in the relative absence of competition from other species. The small rise in water temperatures in the early spring may be advantageous in enabling increased rates of metabolism in company with higher levels of illumination. The picture is complicated by the 'seeding' process which must precede the spring growth. In some years the early growth in coastal waters may be of coastal (neritic) species (e.g. *Skeletonema costatum*); in other years an influx of diatoms from the open sea (due to preceding wind conditions) may form the 'seeding' population. Lake populations of *Asterionella formosa* can only grow from colonies which have survived the winter months, since no 'resting' or perennating stages are known. The marine *Skeletonema costatum* must similarly be 'seeded' from overwintering cells. By contrast, *Melosira italica* in the English lakes is already present in considerable quantity before onset of the outburst. As already noted, its disappearance coincides with the onset of stratification (p. 26).

More subtle factors may also be operative at the commencement of the spring outburst. Following the application of a bioassay technique the condition of the sea prior to the onset of the spring growth has been described as 'poor quality' (JOHNSTON 1963b). It is also suggested that some 'modification' takes place when light and other factors are suitable, and that this change occurs before the spring outburst gets under way. The nature of this change remains a matter of speculation.

The striking seasonal feature of a spring outburst has naturally attracted a great deal of attention, but it is necessary to stress that this is not a feature of all water masses. In some Arctic regions (Fig. 4–1a) the maximum growth of phytoplankton is a summer phenomenon, a short period of very high productivity accompanying the melting of ice and almost continuous day. Growth of phytoplankton will proceed under ice as long as snow cover is minimal. A species succession with the seasons has been described for far northern waters. In Antarctic waters also a seasonal succession has been described with small and rapidly growing diatom species characterizing the spring outburst. The timing of the main diatom outburst in Antarctic waters changes with latitude. At the highest southern latitudes there is a rapid growth of phytoplankton immediately following midsummer (February, March). Farther north the timing of this outburst is progressively earlier and extends over a longer period (November–December and March–April) (Fig. 4–1d,e). By comparison, there is almost continuous production of phytoplankton in tropical waters (Fig. 4–1c) although limited by the well-developed thermocline. Nor is the spring outburst a typical feature of all freshwater lakes. There are numerous examples, not restricted to higher latitudes, where the diatom maximum is delayed until about midsummer, and often this is the sole peak of abundance for the year.

4.3   Summer

In north temperate seas and in many lakes where thermal stratification occurs with the onset of summer, the region of the thermocline becomes an effective lower boundary. This is often a period of small numbers but great species diversity. There will be competition for available nutrients together with grazing pressure from zooplankton, and, particularly in fresh water, the occasional onset of parasitic infections. In north temperate seas the summer floras are notable for their dinoflagellate complements (species of *Ceratium* (Fig. 4–4a,c) and *Peridinium* (Fig. 4–4d,e)) together with diatoms (e.g. *Guinardia flaccida* (Fig. 4–4f) and *Rhizosolenia* spp. (Fig. 4–4g–l)). In lakes there is a marked increase in green algae (desmids, unicellular and colonial flagellated algae, and unicellular and colonial nonflagellated forms). Flagellates of the classes Cryptophyceae and Chrysophyceae are often present in large numbers. Diatoms which were prominent in the spring outburst (*Asterionella, Tabellaria*) do not entirely disappear, and are to be found throughout the summer and autumn. Low nutrient levels measured in the summer are a result of their rapid uptake and recycling in somewhat crowded and competitive conditions. The factor of competition is likely to be of some importance following the spring outburst. The summer conditions enable green algae (and blue-green algae later in the season) to compete more effectively with diatoms which, in turn, are able to grow faster in the colder conditions of early spring. In ponds the summer abundance of green algae and flagellates is also marked, in addition to a number of diatoms.

4.4   Autumn

In many localities there is a second period of phytoplankton abundance after the summer. This phase of plant growth is also accompanied by a 'shock' period (ROUND, 1971). In the sea this is due to the steady lowering and break-up of the thermocline following the increased turbulence of the surface waters. In lakes stratification ends with 'overturn', when the deeper water mixes with the epilimnion. In both cases there is a sudden

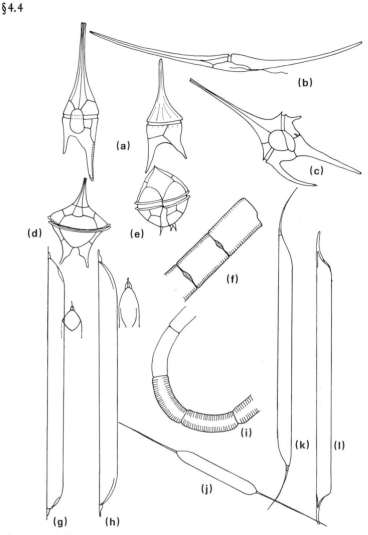

Fig. 4–4 (a) *Ceratium furca* (left) and *Ceratium lineatum* (right). (b) *Ceratium fusus.* (c) *Ceratium hirundinella.* (d) *Peridinium depressum.* (e) *Peridinium pellucidum* (f) *Guinardia flaccida.* (g) *Rhizosolenia shrubsolei.* (h) *Rhizosolenia styliformis.* (i) *Rhizosolenia stolterfothii.* (j) *Rhizosolenia setigera.* (k) *Rhizosolenia hebetata* forma *semispina.* (l) *Rhizosolenia alata.*

temperature change together with an increased supply of nutrients. The phytoplankton organisms now circulate into deeper water and in lower light and temperature conditions. An abrupt change of this magnitude will create stress conditions for some organisms, so that a change from the summer populations is to be expected. The autumnal peak is often numerically of lower order than the spring. In some lakes the water-blooms of blue-green algae are a striking feature of the autumn floras. Whilst the numbers of cells in autumn may well be lower per unit volume than in spring, if the autumn growth is measured in terms of biomass or volume it may equal that of spring. The fall in populations after autumn is due to grazing, greater mixing, reduced light and shorter days.

In the sea a short period of increased diatom numbers is observed. Species of *Chaetoceros, Biddulphia, Rhizosolenia* and *Coscinodiscus* are prominent, together with numerous dinoflagellates. In some localities the silicoflagellates show a small peak. As with the spring floras (p. 54), a listing of autumnal diatoms has been made for north-west European seas, viz: *Bacteriastrum varians, Bellerochea malleus, Chaetoceros affine, Chaetoceros anastomosans, Chaetoceros coronatum, Chaetoceros lauderi, Chaetoceros seiracanthum, Ditylum brightwellii, Rhizosolenia acuminata, Rhizosolenia calcar-avis, Rhizosolenia deliculata, Rhizosolenia fragilissima, Stephanophyxis turris.*

Again, no absolute ruling should be made with these; *Ditylum brightwellii,* for example, is frequently observed in other seasons. Diatoms also increase in numbers in lakes, often with the species prominent in the spring outburst reappearing.

4.5 Winter

In seas of middle latitudes this is the period of sparse phytoplankton populations and low productivity. Species of *Coscinodiscus* and *Biddulphia* are common together with a few dinoflagellates. In coastal waters benthic diatoms living on mud surfaces are carried up into the surface waters by turbulence. In lakes also the populations are sparse, with overwintering cells of species which are prominent in spring (e.g. *Asterionella formosa*) and some desmids. *Melosira italica* occurs in quantity in some English lakes from October onwards (p. 59).

4.6 Interactions

Seasonal changes in phytoplankton involve successions of organisms. Sucessive imprints are made on the water by the dominant representatives, due to factors called ectocrines by LUCAS (1947). These encompass a variety of metabolites, nutrients, hormones and stimulatory and inhibitory compounds. Many of these substances are water soluble. Phytoplankton live and die in a medium which allows rapid interchange of metabolites. Some ectocrines are probably present in quantities which are too small for satisfactory analysis.

4.6.1 Nutrients

PEARSALL's (1932) studies on the English lakes led to the conclusion that four events could be associated with nutrient status, viz. diatom increase with high nutrient status; desmids and other green algae in summer when nutrients are low; numerous blue-green algae when the organic content is high, and the appearance of *Dinobryon* in hard-water lakes when silicate is low and the nitrate:phosphate ratio has increased. *Dinobryon* grows in other lakes under different conditions, however. The spring and autumn diatom maxima, when they occur, commence at times of high nutrient content. In spring this follows the winter mixing of the water, and in autumn both lowering of the thermocline and mixing in the sea and 'overturn' in lakes enrich the surface waters. The summer populations of desmids and green algae in lakes, whilst accompanied by a low calcium content and a low nitrate:phosphate ratio, may also be an expression of the 'opportunist' nature of organisms with the ability to utilize rapidly nutrients set free by recycling in the epilimnion. At the same time, nutrient levels are low and remain so in summer, due to thermal stratification and the reduced input of streams and rivers. Long-term observations (21 years) on desmids in Lake Windermere (CANTER and LUND, 1966) have shown striking correlations between desmid maxima and mean surface temperatures rather than with radiation input and day length.

The association of blue-green algae with organic substances may well be a reflection of their extracellular yield, not dependence on a nutrient source (FOGG, 1963). Organic substances may regulate the ionic environment by complex formation with inorganic nutrients. Nitrogen fixation by filamentous blue-green algae may compensate for nitrate depletion. Whilst some blue-green algae utilize organic substances when grown in the dark, most are photosynthetic.

Nutrient factors and seasonal changes in temperate seas show points of similarity with lakes. Summer populations of diatoms and dinoflagellates grow in a nutrient depleted medium following the spring growths. The summer occurrence of dinoflagellates can be linked with increased temperatures and higher concentrations of soluble organic matter. Some diatoms appear able to thrive with certain nutrients (e.g. silicates) in much lower quantities than required by preceding species. The later species may be thinner-walled, or more successful at metabolizing smaller quantities of silicate at the same time as nitrate and phosphate concentrations are low.

4.6.2   Vitamins

The occurrence of vitamins of the B group in nature has been described (p. 34). Vitamin B_{12} plays an important part in the successions shown by some organisms, particularly where an exogenous source is a necessity.

A dependence of certain diatoms on exogenous sources of vitamin B_{12} has been shown for the nutrient-poor Sargasso Sea. MENZEL and SPAETH (1962), using a bioassay method with the diatom *Cyclotella nana*, found a B_{12} concentration of 0.07–0.1 ng dm^{-3} in April. This maximum was probably due to mixing from deeper water. A diatom bloom (mainly of *Rhizosolenia stolterfothii* and *Bacteriastrum deliculatulum*) followed.

The B_{12} concentration fell to 0.1 ng dm^{-3}, and the short diatom bloom was followed by a dominant growth of *Coccolithus huxleyi* (a coccolith-bearing organism) which is usually present throughout the year. *Coccolithus* is not dependent on exogenous sources of B_{12}. Recent work using bacteria-free cultures have shown that *Coccolithus huxleyi* is not only independent of an outside source of B_{12} but will liberate the vitamin, together with biotin, into the medium, (CARLUCCI and BOWES 1970 *a,b*). But *Coccolithus* is dependent on external sources for vitamin B_1 (thiamine). Three other organisms, the dinoflagellate *Gonyaulax polyedra* and the diatoms *Skeletonema costatum* and *Stephanopyxis turris,* were found to require B_{12}, but were able to liberate thiamine and biotin. The vitamins are produced by healthy cells and released from dead cells. A vitamin-requiring species could then utilize the products of another organism. This has been demonstrated in culture experiments by growing *Skeletonema costatum* (requiring B_{12}) in the presence of *Coccolithus huxleyi* (requiring thiamine) in bacteria-free cultures. These striking experimental results have been linked with seasonal successions (PROVASOLI, 1971). Thus, in the Sargasso Sea the B_{12}-dependent diatoms of the small April bloom could supply thiamine for the subsequent growth of *Coccolithus*. An insufficiency of thiamine can prove limiting to B_{12} production by *Coccolithus*. The slow accumulation of B_{12} in the following seasons may be due to continuing growth of *Coccolithus*. A further correlation with the seasonal successions of north temperate seas may be envisaged, but must at present be regarded with caution. The B_{12} content of the sea is high in winter (p. 34). The spring diatom outburst may then be of organisms requiring B_{12}, enabled to grow rapidly by high concentrations of inorganic nutrients as well as the vitamin, at the same time liberating thiamine and biotin. These would be followed by thiamine-requiring organisms which liberate B_{12}. The summer growth of dinoflagellates would consist mainly of B_{12}-requiring organisms, and the autumn diatom populations might be similarly dependent (PROVASOLI, 1971). It must be repeated that this is speculation; all the links in the chain of events need to be fully investigated by laboratory culture and field observations. The experiments which have shown that marine phytoplankton organisms are capable of both producing and secreting B_{12} are highly significant. For many years it has been assumed that bacteria are the main providers of B_{12} in marine habitats, and that the algae supported bacterial growth by their extracellular products and by providing organic substrates on death. The possibility that widespread liberation of B_{12} can proceed from living algae may well mean a reassessment of B_{12} cycling patterns in sea and freshwater. At present more is known about circulation of the vitamin in marine phytoplankton. Some culture studies with freshwater phytoflagellates have produced evidence of inter-dependence of vitamin secretion and uptake.

The results of experiments on B_{12} production and uptake provide striking verifications of some earlier speculations (e.g. HUTCHINSON, 1944), that algal succession may be significantly influenced by organic growth-promoting or growth-inhibiting substances present in trace quantities.

4.6.3 Other biologically active substances

Much of the evidence that antibiotic substances play a significant part in algal successions has come from culture experiments. Only experiments carried out with algae free from bacterial contaminants are significant. Whilst it can be argued that in nature bacteria are present anyway, it cannot be assumed that bacteria growing with algae in culture flasks resemble those in nature either in numbers or species. LEFEVRE, JACOB and NISBET (1952) gave evidence of antibiotic properties of extracts from cultures of *Scenedesmus quadricauda* and *Pandorina morum* which had a variety of effects on other algae, including growth inhibition and death in some cases and growth promotion in others. PROCTOR (1957) showed that the toxic substance produced by *Chlamydomonas reinhardii* which inhibited the growth of *Haematococcus pluvialis* was a fatty acid which accumulated in the cells and not in the medium, and was released after death. It is therefore possible that dense growths of some phytoplankton organisms influence the succession patterns by the production of inhibitory substances after mass deaths.

Antibiotics which destroy bacteria are produced by some algae. These compounds would influence succession by their effects on bacteria associated with other algae. Compounds which are auxin-like and gibberellin-like in their properties have been isolated from both cultures of phytoplankton organisms and natural waters. These in turn may influence the succession of organisms or of bacteria. Some examples of phytoplankton antibiosis have been described from interactions with zooplankton (see p. 79). In an interesting study in which antimetabolites were tested on both cultures of marine phytoplankton and natural populations it was found that early spring forms (*Skeletonema* and *Thalassiosira*) were more sensitive to antimetabolites than diatoms (*Chaetoceros* and *Rhizosolenia alata*) which appeared later in the succession (JOHNSTON, 1963*a*). Thus the organisms at later stages in the succession may also be those less susceptible to antibiotic activity.

4.7   Bacteria and fungi

Whilst animals and plants outweigh the bacterial floras in aquatic habitats, the significance of bacteria in converting organic substances to inorganic and *vice versa,* and in mineralization and breakdown processes, cannot be overemphasized. Bacteria occur freely suspended in the water, associated with organic debris, attached to plants and animals, and in surface layers of mud and sand on the seafloor or in lakes. They are more numerous in coastal waters than in the open sea. In the oceans they are found in greatest numbers in association with phytoplankton, and close to the seafloor. In lakes they are abundant in the mud and organic debris of the marginal areas and lake floor. They are numerous in lake water, particularly after rainfall. When present in large concentrations bacterial cells are probably grazed by the smaller zooplankton (p. 76). Suspended bacteria utilize organic substrates at great dilutions. At times of 'superfluous feeding' by zooplankton, however, an abundance of nutrient material for bacteria will be excreted into the water. The metabolic activities of aquatic bacteria yield substances of vital importance in nutrient cycles. Oxidation processes convert ammonium compounds to nitrite, nitrite to nitrate, and sulphide to sulphate. The reduction of nitrates to nitrites, sulphates to sulphide and sulphur and both nitrogen fixation and nitrogen liberation, occur in the sea and freshwater. The synthesis of new cell material by bacteria will lead to the incorporation of nitrogen and phosphorus, and these elements will be liberated as ammonium compounds and phosphate after death. Mineralization processes on the seafloor and in lakes will temporarily lock away certain inorganic elements. If aerobic conditions persist on the mud surface at the bottom of a lake the oxidized layer so formed will prevent recycling of certain inorganic nutrients into to photosynthetic zone. With anaerobic (reducing) conditions this recycling process can take place.

Aquatic bacteria are characterized by enzyme systems which appear capable of degrading a wide variety of organic substances. The decomposition of gelatins, protein degradation products, urea, cellulose, chitin, lignin, alginic acid and fats have been described for various heterotrophic saprophytes. In freshwater habitats the fungi play a significant part in recycling processes by decomposing organic matter in the sediments of lakes and rivers. Their role is considered of equal importance to that of bacteria. Less is known about the activities of marine fungi, but their contribution is probably significant.

The sparcity of bacteria in the oceans has been attributed to grazing by zooplankton. Sizeable 'blooms' of bacteria accompany the waning of a phytoplankton population. It is possible that the small numbers of bacteria associated with the waxing phytoplankton populations are held in check by competition for nutrients and by the antibiotic products of diatoms. This small but persistent bacterial population will then be the 'seed' for the bloom when the diatom numbers decline.

4.8 Perennation

Perennation, i.e. the survival of organisms during unfavourable periods, is an important means of maintaining phytoplankton successions. Resting spore formation following loss of sap and rounding-off of cell contents is observed with diatoms in inshore waters (Fig. 4–5a–c). The denser cells sink and germinate only when re-circulated into the upper layers if conditions are favourable. Shortage of nutrients, low levels of illumination and low temperatures are likely factors inducing spore formation. Individual

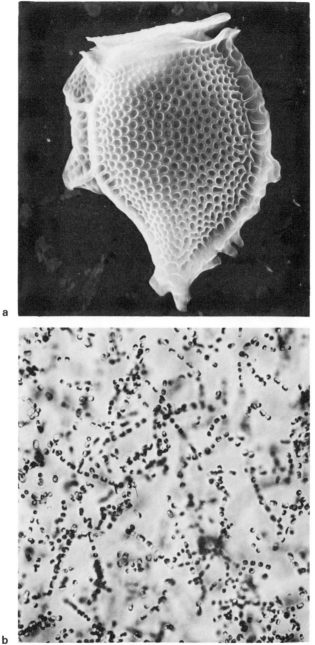

Plate 3 (a) Scanning electron micrograph of the dinoflagellate *Dinophysis norvegica* (× 1500) (courtesy of J. D. Dodge, Birkbeck College, University of London). **(b)** The common neritic diatom *Skeletonema costatum* (courtesy of D. S. Harbour, Marine Biological Laboratory, Plymouth). (Length of individual cells 5–12μm.)

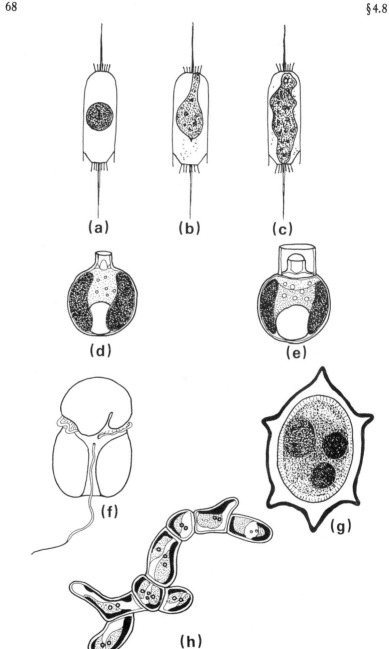

Fig. 4–5 **(a)** *Ditylum brightwellii*—resting spore. **(b–c)** Stages in germination. **(d)** *Dinobryon* cyst. **(e)** *Uroglena* cyst. **(f)** *Wolozynskia tylota*. **(g)** Cyst stage. **(h)** *Apistonema* filament. (**a–c** after GROSS, 1937; **g** after BIBBY and DODGE, 1972.)

cells probably tide a species over unfavourable periods in the open sea without spore formation. Diatom cells can survive winter freezing in colder waters. In freshwater habitats the survival of a species may depend on individual cells (*Asterionella formosa*), or on the occasional formation of resting spores, (e.g. some desmids and colonial green algae). For some organisms the annual production of resting spores is vital for their continued existence (Fig. 4–5d,e), (e.g. *Dinobryon, Uroglena* and dinoflagellates). The spring outburst consists of diatoms which lack a perennating phase. The seasonal appearance of spore-producing algae is always later in the year. Dinoflagellates form thick walled cysts in freshwater habitats. Encystment of *Wolozynskia tylota* commences in May after the spring appearance of motile cells (Fig. 4–5f,g) usually when water temperatures reach 12–15 °C (BIBBY and DODGE, 1972). Cyst formation is accompanied by cell organelle changes indicating reduced metabolism. Perennation and stress tolerance are the least explored aspects of phytoplankton biology.

4.9 Benthic (sedentary) phases

Coccolith-bearing flagellates (e.g. *Cricosphaera* sp. Fig. 1–7) are abundant on occasions in inshore and estuarine waters. In culture they form a filamentous sedentary phase reminiscent of the small plant *Apistonema* found on rocks and wooden structures in inter-tidal areas and estuaries in the region of high water mark (Fig. 4–5h). Isolates of *Apistonema* from nature have produced *Cricosphaera*-like flagellates in culture, and the flagellates in turn have formed the *Apistonema* phase. *Apistonema* filaments withstand extreme stresses and subsequently yield the flagellates on the return of more favourable conditions. The sedentary phases thus serve as a continuing source of coccolith-bearing flagellates.

4.10 Phytoplankton associations

Plankton 'indicators' of water fertility are known. The arrow-worm *Sagitta elegans* with associated organisms in the Western Approaches and English Channel was found to be indicative of productive waters, whilst the related *Sagitta setosa* was found in a much less productive community. Zooplankton can also be grouped in major geographical areas. Certain broad divisions of phytoplankton association have been recognized: Arctic, Temperate, Tropical, Antarctic, Cosmopolitan and Bipolar. Cosmopolitan species include the common *Skeletonema costatum*. Other cosmopolitan species are illustrated in Fig. 4–6a–i. Certain cosmopolitan species may only survive in very small numbers in some regions and make little contribution to overall productivity. Widespread transport of plant cells will occur in the absence of barriers in the oceans. Regional differences in species distributions may be due to the establishment of physiological 'races' or 'strains'.

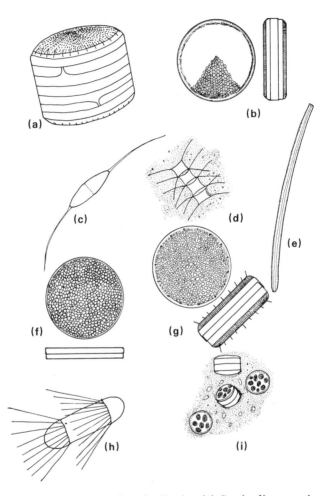

Fig. 4.6 Diatoms of cosmopolitan distribution. **(a)** *Coscinodiscus concinnus*. **(b)** *Coscinodiscus lineatus*—valve and girdle views. **(c)** *Nitzschia closterium*. **(d)** *Chaetoceros sociale*. **(e)** *Thalassiothrix longissima*. **(f)** *Coscinodiscus radiatus*. **(g)** *Coscinodiscus excentricus*. **(h)** *Corethron criophilum*. **(i)** *Thalassiosira subtilis*.

In the much studied North Sea sixteen water masses with associated phytoplankton have been identified, the majority being inshore areas. Inflowing Atlantic water with a sparse flora can also be traced. In the English lakes a number of morphological strains of the diatom *Tabellaria flocculosa* have been described, but these are not restricted to specific lakes. The 'Compound Phytoplankton Index', or 'Compound Quotient',

(NYGAARD 1949), was formulated as a result of the frequent association of numerous desmid species with oligotrophic lakes:

$$\text{Compound Index} = \frac{\text{Number of species of } \textit{Cyanophyceae, Chlorococcales, Centric Diatoms, Euglenoids}}{\textit{Number of species of desmids}}$$

The value of the Index as an indication of nutrient status of a water mass has been questioned, particularly since tow net samples were used. BROOK (1964), by restricting desmid species to those truly planktonic, indicated that reasonable correlations between Compound Index and nutrient status could be obtained for some freshwater lochs in Scotland.

4.11 Palaeolimnology

The sediment profiles of a lake can yield much information on both its history and that of its drainage area. The oldest sediment layers of lakes formed by glacial action consist of pale-coloured inorganic clays. Post-glacial clays are dark in colour due to the organic matter from biological activity in developing lakes. Radiocarbon dating enables a fairly accurate timing of the formation of the various strata. Core samples of lake sediments show stratification through colour, in chemical differences, and in the preserved remains of organisms. Pollen grains from flowering plants in the drainage area offer the most complete record. These are well preserved and identifiable. The frustules of diatoms and cysts of Chrysophyceae, both of silica, are also preserved, and the cyst stages of dinoflagellates. The changing pollen record gives a graphic record of vegetational history and of man's influence on nature. Diatom frustules in different layers may also be indicators of changed environmental conditions.

5 Interactions with Other Organisms

Biochemical analyses of marine phytoplankton have shown that plant tissue is similar to animal tissue in being rich in protein. Their composition show similarities in the amino-acid types and the quantities present. Taurine, found only in animals, is the one exception.

5.1 Zooplankton

Analyses of total particulate matter in sea water and of the diatom *Skeletonema costatum* have shown close similarities in amino-acid composition. Carbohydrates and lipids are present in smaller amounts. Altogether there seems little to choose between various phytoplankton species in terms of their potential food value. Like all foods, however, this is not the whole story. Edibility, digestibility and, with zooplankton, ease of capture must also be considered. Zooplankton feeding on diatoms have also an indigestible silica shell to deal with. Thick-walled green algae (e.g. *Chlorella*) may not be entirely suitable as food. Some phytoplankton organisms appear unacceptable because of the production of antibiotic substances. The easily digested, naked phytoflagellates may be too small to be captured by water-straining mechanisms of the larger zooplankton. Some flagellates are toxic to animals. The larval stages of invertebrates, prominent components of the zooplankton, utilize microbial flagellates as food if available in sufficient quantities. Sessile filter-feeding molluscs will also utilize food organisms in this size range. In the rich diatom growth of the spring outburst certain diatoms may be preferentially eaten. The copepod *Calanus* will feed on the diatom *Lauderia* in preference to *Ditylum*. It is also able to feed at faster rates on larger diatoms than on organisms in the 10 μm size range. Recent studies using *Calanus helgolandicus* and the diatom *Ditylum brightwellii* compared feeding rates with single diatom cells and recently divided cells still joined to one another. Cell division was near synchronous (i.e. a large number of cells were dividing at the same time). Recently-divided and still-joined cells show a volume increase of 50% over single cells. When the divided and joined cells exceeded 40% of the population they were almost the exclusive choice of *Calanus* as food. If 20% or less of the population were paired single cells only were eaten. At times of spring outburst near-synchronous divisions do take place. Whether cells are eaten before or after division will clearly be of ecological significance. Certain flagellates may evoke distinct avoiding actions by zooplankton, whilst others prove attractive. Mysids swim towards water containing the diatoms *Skeletonema, Thalassiosira* or *Bid-dulphia*—genera often abundant in the spring outburst. These animal migrations have been observed with actively growing diatom cultures. Aged cultures might prove less attractive—*Daphnia* feeding was inhibited with senescent cultures of *Chlorella*. This was considered to be due to production of the antibiotic chlorellin. Compounds of this type are formed in spent cultures. Their production by waning plant populations may exert a similar influence on zooplankton in nature. Inhibitory metabolites may form part of a 'conditioning' process, and so be responsible for 'animal exclusion' from areas rich in phytoplankton. Laboratory experiments on choice of food organisms by zooplankton have yielded some unexpected results. It seems odd that the flagellate *Dunaliella viridis* should be more suitable as food for the brine shrimp *Artemia* than the related *Dunaliella salina*. Similarly the freshwater *Scenedesmus spinosis* is a much better food for *Daphnia* than both *Scenedesmus oahuensis* and *S. quadricauda*. Many experiments testing the suitability of phytoplankton organisms as food have been carried out. These experiments have used unialgal or mixed cultures, so that the condition of the food organism may not have been the same as in nature. Under laboratory conditions *Calanus finmarchicus* was found to eat all types of diatoms presented. Old diatom cultures were less well digested than new. Dinoflagellates on the whole were well digested. The flagellate *Prymnesium parvum* was also eaten, and this organism is known to be toxic to fish (p. 79). Most green phytoflagellates were eaten and digested. *Chlorella* appeared to resist digestion and viable unaltered cells were isolated from the *Calanus* faecal pellets.

The food requirements of zooplankton will differ with their size and longevity. Male *Calanus* live for shorter periods and eat less than female *Calanus*, which also show rates of egg production closely related to amounts of food eaten. Starved animals produce few eggs on which they feed if plant cells are insufficient. Turnover of food to eggs is rapid. Experiments with ^{32}P labelled phytoplankton have shown a 70% concentration in *Calanus* eggs after 5–6 days. Larval stages of *Calanus* and of the common intertidal barnacles show varying requirements in terms of both the sizes and types of phytoplankton cells eaten.

5.1.1 Superfluous feeding

The rate of faecal pellet production is a quick measure of zooplankton feeding in both laboratory and sea. In a rich diatom suspension *Calanus* will feed rapidly and yield faecal pellets containing undigested and partially digested plant cells. This voracious habit has led to the concept of superfluous feeding, i.e., the useless destruction of algal cells. In such conditions the *Calanus* maintains a maximum respiratory rate and egg production without expenditure of energy in seeking food. Estimates have placed the wastage rate as two-thirds of the algal production but the abundance of food means that the animals are well fed despite this wastage. Algae labelled with ^{32}P yield an appreciable quantity of the isotope to the gut wall of *Calanus* on digestion—even when passed through rapidly. It is also known that faecal pellets in nature have a very low phosphorus content suggesting an appreciable uptake during digestion. Further, partial digestion of diatoms, including fracture of the silica walls, allow a more rapid recycling of silica. The particulate organic matter voided will serve as food of organisms in deeper water. Zooplankton in the euphotic zone may also feed on it when deprived of plant cells. Superfluous feeding and production of organic matter at times of phytoplankton abundance may well be an essential step in making food available to benthic animals in deeper water.

5.1.2 Grazing and exclusion

GRAZING. The inverse quantitative relationship between phytoplankton and zooplankton in temperate seas is well known. Whilst distributions in nature are patchy even at the time of the spring outburst, given suitable conditions the increase in numbers of phytoplankton will far exceed the rate of

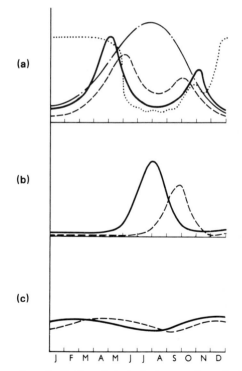

Fig. 5–1 Seasonal productivities of phytoplankton and zooplankton. **(a)** North temperate seas. **(b)** Arctic sea. **(c)** Tropical seas.

——————— phytoplankton   • • • • • • • • • plant nutrients

– – – – – zooplankton   —— · —— · light intensity

zooplankton grazing. The magnitude of the spring peak of phytoplankton numbers illustrates the initial lack of herbivore grazing pressure. Due to the different feeding requirements of larval stages there is some delay before herbivore grazing pressure exceeds the rate of plant cell increase, so reducing plant cell numbers before exhaustion of the basic nutrients (Fig. 5–1a). At no time is there an equilibrium between phytoplankton and zooplankton. Rates of plant cell production will vary with time; herbivorous zooplankton will be the prey of carnivores in turn allowing temporary increases of phytoplankton. Superfluous feeding enables recycling of mineral nutrients from the organic matter formed. In arctic waters the midsummer phytoplankton peak is reduced by grazing, (Fig. 5–1b) but there is a delay. In tropical waters production of plant cells is almost continuous. Grazing by herbivores is steady throughout the year (Fig. 5–1c). There is a shorter delay period between phytoplankton increase and zooplankton egg production.

EXCLUSION. The inverse phytoplankton:zooplankton relationship has also been interpreted as animal exclusion. Phytoplankton metabolism will condition the water, and the 'exclusion' hypothesis proposes that animals avoid areas rich in phytoplankton because of some toxic or unpleasant factor; in nature it is known that dense concentrations of some algae are avoided, e.g. *Phaeocystis* (p. 79), *Coscinodiscus* and *Rhizosolenia*. Experiments show that animals will swim towards dense patches of some algae, but that aged cultures (the equivalent of waning populations) are avoided. Toxic flagellate blooms (p. 80) will undoubtedly cause exclusion.

5.1.3 Grazing or exclusion?

The weight of evidence from both laboratory and field observations indicates that grazing effects are sufficient to reduce plant cell numbers following a spring outburst. BAINBRIDGE (1953) has proposed that the relationship between plant and animal increase results from a dynamic inter-relationship of growth, grazing and migration. A dense patch of phytoplankton develops under favourable conditions as a result of a 'seeding-in' process, (Fig. 5–2a), and attracts zooplankton. Initially plant growth is unaffected by grazing, but the attraction of more zooplankton exerts significant grazing pressure (Fig. 5–2b). Clearance of phytoplankton leaves a growing zooplankton patch with the neighbouring sea free of zooplankton wherein phytoplankton growth again starts (Fig. 5–2c,d). CUSHING (1964) traced the history of a grazing *Calanus* patch in the North Sea for two and a half months in 1954. We have insufficient field data at present to evaluate the exclusion hypothesis.

5.1.4 Synchronization processes

Barnacles have pelagic larval stages prior to settlement. In the Firth of Clyde evidence has been obtained of synchronization between the spring phytoplankton outburst and release of barnacle larvae. Initiation of the release of barnacle larvae was seen to coincide with the spring growth of

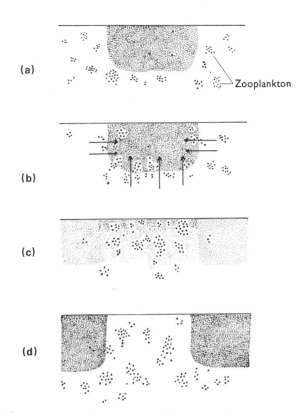

Fig. 5–2 Diagrammatic summary of phytoplankton–zooplankton inter-relationships. **(a)** Inverse relationship at start. **(b)** Commencement of zooplankton migration with some grazing. **(c)** Heavy grazing on completion of migration; growth of phytoplankton on either side. **(d)** Return to inverse relationship. Density of phytoplankton indicated by degree of stippling.

Skeletonema costatum over several years, and to continue over the main period of diatom growth. The numbers of barnacle larvae fell appreciably in years when *Skeletonema* failed to appear although other diatoms (*Chaetoceros* and *Coscinodiscus*) were abundant. Similar processes of synchrony have been observed in Arctic waters coincident with ice melting. Release of larvae appears to be controlled by some internal factor.

5.1.5 Freshwater zooplankton

The principal primary consumers are rotifers and small crustaceans. These feed on small diatoms, organisms of nanoplankton size, organic matter and bacteria. Spring outbursts and declines have been studied mainly with larger diatoms and dinoflagellates which are not subject to predation by rotifers, etc. Spring populations of *Asterionella* in Lake

Windermere are not reduced by zooplankton (LUND, 1965). Protozoa play some part in secondary production (CANTER and LUND, 1968). They prey on colonial green algae and by ingesting whole cells can significantly reduce populations during summer. Partially destroyed colonies and mucilage envelopes will serve as substrates for bacterial activity, and all serve as food for small crustaceans and rotifers. Phytoplankton waning in lakes in early summer could well be the combined result of nutrient shortages, overcrowding, self-shading and onset of thermal stratification. Death and decay of larger algal cells will trigger-off short term outbursts of nanoplankton and encourage growth of protozoa and algae, and subsequently affect later zooplankton populations. Whilst variations in zooplankton abundance are apparently produced by earlier phytoplankton growths, evidence for direct interactions is less clear cut—a paradox noted by HUTCHINSON (1967).

5.2 Food chains, food webs and trophic levels

The sequence Diatom → Copepod → Herring is a simple food chain. A more complex food web culminating in the adult herring occurs in nature (see PHILLIPSON, *Ecological Energetics*, p. 9). The herring stands at the summit of a pyramid of numbers, the base line being the large number of supporting primary producers. Successively the numbers of primary, secondary and tertiary consumers gets smaller and smaller. At each level some of the organic matter consumed is utilized in respiration as well as formation of animal tissue, (Fig. 5–3). If the successive steps in a food

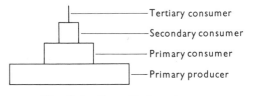

Fig. 5–3 Pyramid of numbers.

chain are interpreted in terms of energy transfer, estimates have shown that herbivore utilization of plant organic matter may be as low as one-tenth, and will not exceed one-third. Hence at the apex of the pyramid the ultimate production of animal flesh will be but a minute fraction of the original plant photosynthesis.

5.3 Organic detritus

An appreciable quantity of the particulate matter in the sea is present as non-living detritus. This outweighs the living phytoplankton, sometimes by a tenfold factor. The phytoplankton appears to be the main contributor to this particulate matter. The quantities of organic matter appear to follow the changing populations of phytoplankton with the seasons, being very high in spring and low in summer. The large quantities of organic particulate matter in the winter months (when phytoplankton is sparse) may result from precipitation of dissolved organic material due to turbulent water conditions. The part played by this organic matter in the economy of the sea is not yet determined. Experimental studies with *Calanus* in which organic detritus was used as food proved inconclusive despite its known protein and carbohydrate content. It is possible that organic detritus is eaten with phytoplankton if aggregates are formed in nature. If organic detritus of plant origin is utilizable as supplementary food it offers a further dimension to the initial plant production.

5.4   Phytoplankton and shellfish larvae

Commercial shellfish (clams and oysters) can be reared through the critical larval stages under controlled conditions. The filter-feeding larvae can be reared on a variety of phytoplankton organisms. Natural sea water enriched with commercial fertilizers produce mainly *Chlorella*-like cells and flagellates. Mass cultures of flagellates like *Monochrysis lutheri* and *Isochrysis galbana* have been used successfully. Colourless flagellates and organic detritus are of lower nutritional value than photosynthetic organisms. In nature high larval mortality rates occur if their release coincides with an abundance of a toxic flagellate (e.g. *Prymnesium parvum*, p. 79).

5.5   Extracellular products

Extracellular products are compounds produced by healthy, actively growing algae which pass from the cell into the medium. The compounds are not to be confused with the substances yielded by cell decomposition or autolysis. The variety of substances liberated is considerable, including organic acids (particularly glycollic acid), carbohydrates, amino-acids and peptides, vitamins, growth substances, antibiotics, enzymes and toxic compounds. The quantities released are also variable—1.4 mg of glycollic acid per dm^3 has been measured in the inshore waters of Washington State, USA. In Lake Windermere quantities up to 0.47 mg dm^{-3} have been found. A high proportion of nitrogen fixed by planktonic blue-green algae is released in a combined form. Certain extracellular products have markedly toxic effects on other organisms. The spectacular red-tide phenomenon is linked with such effects (p. 80). Production of external metabolites imprints a biological history on a water mass which will influence organisms next in succession. Substances of this nature, called 'ectocrines' (p. 62), have been the subject of much investigation and speculation.

5.6   Antagonistic reactions and antibiosis

Antagonistic reactions are due either to the organism or to metabolic products in the medium. Antibiosis is the reaction when metabolic products are involved.

5.6.1   Fish kills due to Prymnesium parvum

Whilst isolated reports of fish kills due to the flagellate *Prymnesium parvum* have come from several parts of Europe, the most disastrous effects were obtained in highly eutrophic brackish-water Israeli fish ponds. Fish poisoning involves the central nervous system, and toxin stability was found to be dependent on a number of factors in the water (pH, suspended soil particles, phosphate, ultra-violet irradiation). Certain 'strains' of flagellate may synthesize the toxic substance. Samples containing 240 cells mm³ were far more toxic than another sample with five times as many cells. Certain co-factors (divalent ions and organic compounds) appear necessary before maximal toxic effects are obtained. The toxic material has been isolated (prymnesin) and is described as proteinaceous. Addition of ammonium sulphate to the ponds has been found to be an effective control measure.

5.6.2   Phaeocystis pouchettii

Colonies of cells (in vast numbers in mucilaginous spheres up to several centimetres in diameter) are abundant in north temperate inshore waters in late spring and the summer. Cells released as individual flagellates can produce dense populations of between $28–193 \times 10^6$ dm^{-3} (e.g. as in the Irish Sea). The vast numbers of colonies colour the sea yellow-brown (a fisherman's sign called 'weedy water'), and fish avoid these patches. A *Phaeocystis* patch over 100 miles in extent was observed in the North Sea in 1927. Patches of this size have caused deflections of herring migratory routes, with shifts of spawning grounds and cutting-back of fisheries. Zooplankton usually avoid this species, which probably accounts for the rapid build-up of populations. In Antarctic seas the crustacean *Euphausia superba* (krill) feeds on *Phaeocystis,* and accumulates acrylic acid formed by breakdown of the dimethyl-β-propiothetin present in the alga. Penguins feeding on krill acquire a gut antibiosis due to the antibacterial activity of acrylic acid. Seagulls feeding on the blue-green alga *Trichodesmium erythraeum* in tropical waters acquire gut antibiosis due to another (unknown) compound.

5.6.3   Blue-green algae

Seasonal 'blooms' of colonial and filamentous blue-green algae commonly occur in lakes. Certain of these have been found to be toxic to water-fowl and domestic animals. Filamentous representatives (*Anabaena flos-aquae, Aphanizomenon flos-aquae*) and colonies of *Microcystis aeruginosa, Coelosphaerium kutzingianum* and *Gleotrichia echinulata* are described as toxigenic. The 'strains' which are toxic are morphologically identical with those non-toxic. Polypeptide substances with toxic properties have been identified in *Microcystis* and *Anabaena* (GORHAM, 1964). The bloom-forming *Trichodesmium erythraeum* is also occasionally toxic in tropical waters.

5.6.4 Red tides

The first documented red tide is described in Exodus 7, verses 20; 21, as one of the plagues of Egypt. These phenomena are mainly blooms of toxic dinoflagellates, and can colour the sea red, pink, brown, yellow and green. Certain dinoflagellates have been described as causative agents (*Gonyaulax catenella, G. tamarensis* and *Gymnodinium brevis*). Massive fish kills, contamination of shellfish with consequent death of sea birds, and general loss of amenity in holiday centres can result. Consumption of con-taminated shellfish by man can lead to vomiting, paralysis of facial muscles and limbs, and death from respiratory failure. Bioassay techniques and the routine examination of phytoplankton samples allow some predicting of the times of the year when blooms of toxic potential are likely to arise. Sudden surges in nutrient levels in coastal waters appear to be frequent causes. Upwelling, rivers in flood and domestic pollutants are likely origins of these surges. Mild red tides have been reported recently from the Moray Firth (September 1963) and the Northumberland coast (August 1968).

5.7 Fungal parasites

Evidence that phytoplankton populations are subject to fungal infections has come mainly from studies on freshwater lakes. Both diatoms and desmids have been observed as host organisms. Fungal parasites are found mainly in eutrophic lakes. They are all simple fungi belonging to the Phycomycetes and to the orders Chytridiales, Lagenidiales and Saprolegniales.

The chytrid *Rhizophidium planktonicum* is a parasite in *Asterionella* (CANTER and LUND, 1948, 1951). The parasitic infection can bring about death of the cells and is noticeably of greater severity in late summer and early autumn, sometimes extending into winter. The onset of the following spring outburst may well be delayed and the algal successions changed by the fungal infections of earlier seasons. Other common diatoms suffer similarly from fungal infections. Whilst desmids may also show high levels of infestation with consequent reduction in population, recovery processes appear to be more rapid.

5.8 Symbiosis

Symbiotic associations between unicellular algae and animals are com-mon. Blue-green symbionts (cyanellae) are associated with certain protozoa and algae. Zoochlorellae (green) are found in freshwater animals

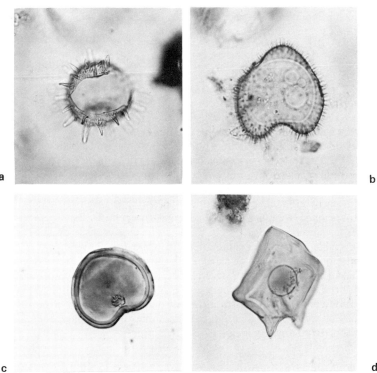

a

b

c

d

Plate 4 Dinoflagellate cysts from marine deposits. (a) Cyst of *Gonyaulax polyedra*; (b) Cyst of *Peridinium claudicans*; (c) Cyst of *Peridinium oblongum*; (d) Cyst of *Peridinium leonis*. (Courtesy of P. C. Reid, Institute of Marine Environmental Research.)

and zooxanthellae (yellow) in marine organisms. Protozoa, sponges, flatworms, hydroids, sea anemones and molluscs are more commonly associated, with the algal symbiont being intracellular in most cases. Transference of symbiont often occurs at the time of reproduction by the animal, but frequently each new animal generation needs to be re-infected with algae. Some regulation of the plant cell numbers is effected by the animal. The algae supply oxygen and soluble carbohydrates and in return receive protection, and carbon dioxide, nitrogen and phosphorus from the animal's excretory products.

5.9   Dispersal by animals

Dispersal and transport of planktonic algae by animals is one way by which species can be introduced into a water mass. Airborne dispersal offers an alternative method. Considerable numbers of airborne algal cells have been detected, but these are predominantly of blue-green algae and unicellular green algae. Insects, fish, water-fowl and aquatic mammals have been examined as possible transporters of viable algae. Living algae have been recovered from both the external surfaces and guts of insects; *Chlorella* sp., *Chlamydomonas* sp., *Chlorococcum* sp. and filamentous forms were isolated. Some desmid species have also been identified in cultures made from insect-gut 'washings'. Insect flight distances of up to 8 miles have been recorded. Hence there can be passive dispersal of algae over appreciable distances. Examinations of water-fowl have shown that the greater quantities of viable algae are borne on the feet. Cultures from bird guts and from droppings of several species of water-fowl show that viable benthic diatoms and filamentous green algae are present. Plankton diatoms are rarely observed. *Melosira* is the one diatom recoverable alive from the faeces of water-fowl. The larger planktonic diatoms appear not to be transported to any significant extent by animals.

5.10   Man and allergies

Allergenic properties have been attributed to some airborne algae—mainly unicellular green algae from soil and aquatic habitats. Diatom growths on the ropes of lobster pots have been shown to be a likely cause of a skin disorder on the hands of fisherman. Respiratory distress and eye discomforts have been ascribed to airborne spray containing dinoflagellates from red tides.

6 Measuring Phytoplankton Populations and Primary Productivity

Phytoplankton populations are measured either by number or the biomass of organisms per measured volume of water. Overall volumes can be determined from the estimates of numbers. Differences between numbers and biomass are indicated in Table 6.

Table 6 Numbers of organisms and biomass determination for phytoplankton from the Gulf of California. (From ZEITSCHEL (1970). *Marine Biology*, **7**, 305)

	% by number	% of biomass (as phytoplankton carbon)
Flagellates $< 5 \mu$m	72	10
Dinoflagellates	13	29
Diatoms	10	51
Coccolithophorids	5	10

6.1 Collecting by nets

The quickest way of obtaining a concentrated sample of phytoplankton is to tow a cone-shaped net (of bolting cloth, silk or monofilament nylon) through the water. The wider end of the net is kept open by a metal hoop, and this is attached to the towrope by a rope bridle (Fig. 6–1a). The narrow end is closed by a metal or plastic receiving vessel (the 'bucket'). When towed through the water a back-pressure builds up at the opening which prevents some water flowing through the net. A fine-meshed net hauled too

Table 7 Mesh sizes of plankton nets

Meshes per inch	Inside measurement of mesh (mm)	
60	0.33	Fish eggs, large zooplankton
100	0.168 ⎫	Zooplankton, large phytoplankton
150	0.094 ⎭	
180	0.079	Zooplankton, phytoplankton
200	0.054	Phytoplankton, small zooplankton

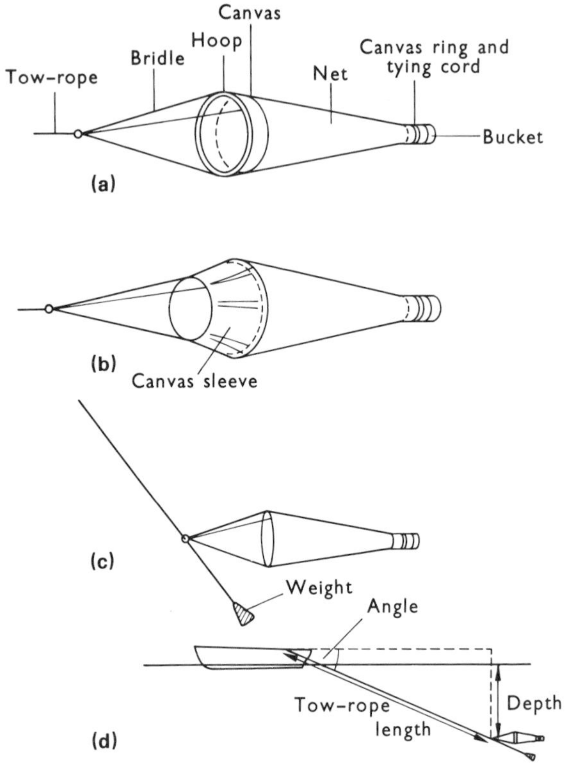

Fig. 6–1 **(a)** Net with rope bridle. **(b)** Net with canvas sleeve. **(c)** Arrangement for sampling in deeper water. **(d)** Method of determining approximate depth of net.

rapidly through the water will produce enough turbulence to deflect the plankton. A tapering canvas sleeve allows more effective filtering by reducing the volume of water entering the net (Fig. 6–1b). A coarse net ensures a fast flow suitable for collecting the larger zooplankton. Slower filtration rates are obtained with nets of finer mesh sizes, and smaller organisms are then collected, (Table 7). Net samples can be collected from various depths (Fig. 6–1c,d), and vertical hauls made between levels with the aid of a throttling device around the canvas opening. The throttling device is triggered by a metal weight or messenger sent down the cable from the boat. Net samples are inadequate as the basis of quantitative studies due to the uncertain nature of the volumes of water filtered and the selective nature of the sampling. Use of the net results in some clogging of the holes so that the rate of filtering undergoes change. Flow meters, (multi-bladed propellers with a counter for logging total revolutions), give more accurate information on the quantites of water flowing through the net, but the selective nature of the method of capturing organisms can exclude the nanoplankton. In oligotrophic lakes with sparse phytoplankton floras net sampling is often the only way of recording the population despite quantitative limitations.

6.2   Tubes and water bottles

In calm conditions a length of hosepipe (up to 5 m) is weighted at one end and this when lowered into the water encloses a known volume of sample. With the upper end closed at the water surface the lower (weighted) end is hauled aboard by means of an attached length of cord and the water sample plus phytoplankton is transferred to a clean container.

A weighted glass or plastic bottle of known capacity and sealed with a rubber bung can be lowered to a required depth in the water. The bung is fixed to a length of stout line and is removed at the required depth. Fixed volume samples of more sophisticated design are used for deeper waters.

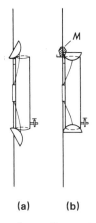

(a)   **(b)**

Fig. 6–2   Van Dorn type of sampling apparatus. **(a)** Open. **(b)** Closed by messenger (*M*).

The Van Dorn bottle (Fig. 6–2) is an open cylinder of known capacity that is let down into the water and automatically closed at both ends by a metal weight or 'messenger' which slides down the cable. The enclosed water is under pressure so that more water cannot enter at other levels during passage to the surface.

6.3   Pumps

A suction pump with a weighted tube of required length can be used to collect plankton organisms at successive levels throughout a water column.

6.4   Preservation of samples

A 10% neutral formalin solution can be used although this is not very satisfactory for delicate organisms. Lugol's iodine solution, (10 g or iodine and 20 g of potassium iodide in 200 cm^3 of distilled water, with 20 g of glacial acetic acid added 2–3 days before use), gives better preservation of

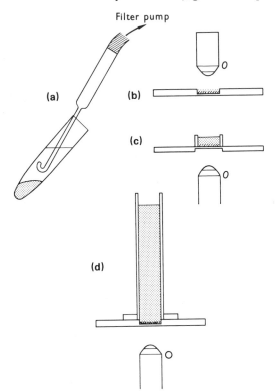

Fig. 6–3   (a) Pipetting away water after centrifugation. **(b)** Shallow counting chamber for use with normal microscope (O = objective). **(c)** Counting chamber for use with inverted microscope. **(d)** Combined sedimenting tube and counting chamber.

flagellates. Lugol's iodine is added to water samples in a ratio of 1 volume of iodine solution to 100 volumes of water. Rates of sedimentation are important here. The time required can be estimated as $T = 3h$, where T = sedimentation times (hours) and h = height (cm) of sedimentation tube.

6.5   Counting: direct counts of living phytoplankton

Concentration of the organisms is usually necessary prior to counting. When the phytoplankton contains large numbers of flagellates çentrifugation is the only reliable method. One disadvantage is that some resuspension of the material occurs when the centrifuge stops rotating. Volumes (10–50 cm³) are centrifuged in graduated tubes at 1500 rpm for up to 30 minutes. The supernatant water is removed without disturbance of the pellet (Fig. 6–3a) until the volume is reduced to 1/10, 1/20 or 1/40 of the original. The plankton 'pellet' is then thoroughly mixed with the volume of water remaining and transferred to a counting chamber. Plankton can be concentrated by the simple device of placing the water sample in a cylindrical plastic container and allowing a filter paper of similar diameter to sink down under the weight of a metal ring. The water can be siphoned off and the concentrated phytoplankton counted. Counting chambers are of various design depending on whether an inverted microscope or an ordinary microscope is available (Fig. 6–3b,c). A moving stage is essential. Haemocytometers can be used as counting chambers if large planktonic algae are absent. Counting chambers combined with sedimentation tubes require inverted microscopes (Fig. 6–3d).

6.5.1   Direct counts—preserved material

When phytoplankton is abundant volumes up to 25 cm³ can be placed directly in the sedimentation tube above the counting chamber and treated with Lugol's iodine solution. When large volumes of water are used sedimentation is first carried out in a large cylinder. A smaller volume of water plus sedimented algae is then transferred to the combined tube and counting chamber.

6.5.2   Counting methods

Whilst centrifuging and sedimentation are carried out with known volumes of water, enumeration of the organisms is made per unit area of the floor of the counting chamber. Hence the importance of not disturbing the sample once settlement on the floor of the counting chamber is completed. The number of organisms counted can then be equated with the original volumes of water used. The counting chamber is first scanned under a low-power lens. Large organisms present in quantity (colonies of blue-green algae, filaments of algae, large diatoms) should be counted for the whole sample. For much smaller organisms transects of the chamber are made with a high-power lens. Two diameter transects which cross are first made, and the area of each transect must be known. This can be determined by using a calibrated eyepiece grid. Two parallel hairs glued to the underside of an eye-piece lens will allow a precise measurement of the area of a field of view if their distance apart is calibrated. The area enclosed by the eyepiece grid or the two hair lines is then examined. Positive guidelines such as these will ensure that organisms are only counted once during the traverse. The overall transect area is the sum of the number of fields counted. The numbers of organisms in the area of the floor of the chamber can be calculated from the total organisms per transect area and the ratio of transect area to chamber floor area. This number can then be expressed per volume of water used for sedimentation.

The problems associated with these counting methods must not be underestimated. Samples must be throughly mixed prior to sedimentation. The water should come to room temperature before treatment, otherwise bubbles will collect and prevent uniform settlement of the organisms. Certain extremely buoyant algae do not settle after fixation with iodine solution, and are poured away when the supernatant water is decanted. Volumes of fresh material counted in small chambers (Fig. 6–3b) will be required if these buoyant organisms are present in quantity. Counting all the organisms in transects of known area assumes an homogenous distribution on the floor of the counting chamber. This is an assumption which needs checking, particularly when large volumes of water are used. A number of transects will need to be counted. Just how reliable a single transect count is in terms of the overall population can be assessed by comparing the numbers of organisms in each of a series of transects. With dense populations of phytoplankton (as from eutrophic lakes), a series of counting chambers of different capacity should be used (e.g. 50, 10 and 1 cm^3; UTERMÖHL 1958). The statistical basis of counting methods are described in a number of papers; those by LUND (1951), LUND, KIPLING and LE CREN (1958); UTERMÖHL (1931, 1958); LOVEGROVE (1961) can be consulted.

6.5.3   Identification

It is difficult to count and at the same time try to identify organisms with keys and illustrations. Recognition of species is a matter of experience. It is advisable to examine the material before counting, and to identify as many of the organisms as possible. It is also useful to prepare a series of pencil or ink drawings on postcards of the species observed. Pinned to a board and placed in proximity to the microscope, these drawings will help to speed up the counting process until one is experienced in recognition.

6.5.4   Membrane filters

A volume of water sample is passed through a membrane filter (e.g. Millipore 0.45 μm pore size). Filtration is carried out with a standard apparatus linked to a filter pump (Fig. 6–4). The volume of water to be filtered depends on the quantities of suspended material. All the phytoplankton organisms are retained, together with fine particulate detritus. Too much detritus will clog the filter; a membrane of larger pore size (1 μm) is then used. To count the phytoplankton organisms the membrane must then be 'cleared'. It is recommended that freshwater samples are washed with distilled water before treatment. Marine residues are washed with a series of diluted seawater samples (90%, 75%, 50%, 25%, 10%, 5%), and finally with distilled water. Washing should be carried out with the membrane still in the filtering apparatus. Subsequent treatments vary. The method of HOLMES (1962) is to dehydrate the residue whilst still on the filtering apparatus by passing through small volumes (10–15 cm^3) of

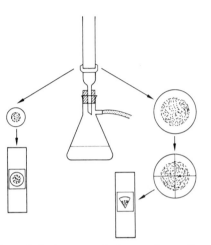

Fig. 6.4 Separation of plankton sample by membrane filter, with subsequent treatments—direct mounting in immersion oil (left) and mounting in Canada balsam after treatment with Fast Green (right).

alcohol of increasing concentrations (10%—absolute alcohol). The sample is stained with Fast Green (0.1% in 95% alcohol) for 20 min then washed again with absolute alcohol. Final clearing is effected in creosote oil, anisole or immersion oil. All phytoplankton cells are stained blue-green or green, and the cleared membrane is dried and mounted on a slide in a suitable medium (Canada balsam in an organic solvent, or Euparal). Membranes of 15 mm diameter can be mounted directly. Larger membranes will have to be cut into sections. MOORE's method (1963) is to apply the immersion oil after washing and without fixing or staining. The oil is applied to the membrane on a slide, left for 24 hours in the dark, and then covered directly with a cover glass. The oil can be sealed by 'ringing' the cover glass with a transparent nail varnish. This second method results in loss of shape of the more delicate organisms, but diatoms and dinoflagellates preserve well. Without staining, the chloroplasts retain a green colour after the death of cells in immersion oil. The stained membrane filters can be kept for several years as long as complete dehydration and clearance have been carried out. Filters mounted directly in immersion oil can be kept for up to 18 months. These methods allow time to elapse between collection and counting. Counting organisms on the filters is subject to the same statistical requirements and principles as counting the sedimented samples.

6.5.5   Use of fluorescent microscopy (WOOD, 1955; 1962)

This method allows a quick discrimination between living and dead phytoplankton. Samples are first concentrated by centrifugation, then examined in a counting chamber with a base of especially thin glass. Using a monocular or binocular microscope with a bright incandescent light and a B 12 substage filter (maximum transmission around 450 nm) and a condenser immersed in parafin oil to concentrate the light, the sample is examined with a OGL (yellow) filter in the eyepiece. Chlorophyll fluoresces bright red. Staining in weak acridine orange (1 part/5000) will give further verification of living and dead cells, particularly if considerable amounts of organic detritus are present. An advantage of this method is that a quick change to normal-light microscopy is possible to verify any doubtful identifications.

6.5.6 Nanoplankton

The sizes and fragility of nanoplankton organisms present particular problems of counting. Their patchy distributions similarly give problems of sampling. Separation of the nanoplankton using a continuous centrifuge has been successfully applied. This facilitates the concentration of organisms from large volumes of water. Even the most delicate flagellates appear able to survive contrifugation at 15 000 rpm, or over. The organisms can then be counted in a suitable chamber. An alternative method involves passing a small volume (50 cm^3) of water through a membrane filter, and then resuspending the filtered material in 1 cm^3 of the water which has passed through the filter. The flagellates in the concentrated sample are then counted in aliquots spread on a haemocytometer slide. Culture techniques have also been applied in quantitative studies on nanoplankton. These are open to the objection that the composition of a specific culture medium may encourage the growth only of certain organisms.

Measurements by direct counts of nanoplankton remains one of the big problems in estimating phytoplankton populations. Continued work is necessary on suitable preservatives and on counting methods. Methods of centrifugation on raw samples have given the best results to date (BALLANTINE, 1953).

6.6. Chlorophyll measurements

Algal abundance can be assessed by extracting chlorophylls and carotenoids in organic solvents, (80% or 90% acetone; 90% or 100% methanol). The water is first filtered through a coarse net to remove zooplankton, then filtered through a 7 cm glass-fibre filter pad previously treated with a magnesium carbonate suspension which assists both in retaining particulate matter and counteracting any acidity that may develop during extraction. Pigment degradation occurs in the presence of acids. Extraction is for 24 hours in the dark at 0 °C or 1 °C, or for a shorter time if the residue is ground in a suitable mill. All debris is removed by centrifugation. The quantity of water to be filtered depends on the amount of suspended material. In a rich eutrophic lake in spring less than 100 cm^3 will be enough; for coastal waters in mid-winter up to 1 litre will be necessary.

Quantities of more than 2 litres will present filtration difficulties. The absorbance of the extracted pigment is measured in a spectrophotometer. Chlorophyll *a,* the pigment universally present in plants, is often used as a standard for measuring plant populations. Other chlorophylls (*b* and *c*), and carotenoids, can be estimated using suitable formulae.

Numerous formulae are available for calculating the chlorophyll *a* content of the water sample: sources such as TALLING and DRIVER, (1963); STRICKLAND and PARSONS, (1968); TALLING in VOLLENWEIDER (1969), should be consulted for the theoretical reasoning behind them.

The equation for calculating the quantity of pigment is

$$\mu\text{g pigment dm}^{-3} = \frac{C}{V}$$

where V = volume of water filtered (dm^3)
and C is calculated from the following formula

$$C \text{ (Chlorophyll } a) = 11.6_{E_{665}} - 1.31_{E_{645}} - 0.14_{E_{630}}$$

Where E_{665} = the extinction (= optical density or absorbency) at 665 nm
E_{645} = the extinction measured at 645 nm
E_{630} = the extinction measured at 630 nm

6.6.1   Plant pigment units (PPU)

The methods described assume a spectrophotometer is available. If not, relative quantities of plant pigment can be measured using a plant-pigment unit (HARVEY, 1934). A standard 'green' solution is used as a basis for comparative measurements. One plant pigment unit is the colour in a solution containing 430 μg of nickel sulphate and 25 μg of potassium chromate dissolved in a known volume of water made slightly acid. 80% acetone is used to dissolve the pigment. A series of tubes is prepared containing an increasing number of plant pigment units (e.g. 2–20 etc.) per cm^3. The colour of the acetone extract is then matched with the range of colours in the tubes. In the original method the plant pigment was measured per m^3 of sea water. The method is applicable to any volume of water filtered, but whilst comparisons between samples can be made in terms of colour density no precise data are obtainable on the quantities of pigment present.

$$\text{Plant pigment units/m}^3 = \frac{1000}{V} \times a \times p$$

V = volume of water filtered (dm^3)

a = volume of acetone extract

p = number of plant pigment units per cm^3.

Arguments have been advanced for and against chlorophyll measurements. The uncertainty that all the pigments in a sample have been extracted and the presence of higher plant debris in the water have been sited as possible sources of error. Further, chlorophyll estimations will often fail to reflect either the numbers or types of organisms present. On the other hand, given particular seasonal conditions (e.g. the almost uni-algal growth obtained in the early stages of some spring outbursts), pigment measurements will strikingly illustrate the waxing and waning of populations.

6.7   Electronic counting

Electronic particle counting techniques originally designed for routine counting of blood cells have since been applied to other problems in the biological sciences and to industry. The Coulter Counter is the model most used in plankton studies. The principle of the instrument is based on the changes which take place when a particle moves into an electric field. If the resistivity of the particle differs from that of the electrolyte, its entry into the field will cause a change in electrical properties. The particle will also displace its own volume of electrolyte and the accompanying change in electrical properties will be directly proportional to particle volume. The electrical disturbances caused by the particles when they are made to pass through a small aperture are automatically measured and counted. The instrument is solely a particle counter and cannot differentiate between organic detritus, inorganic material, and dead and living diatom cells. Parallel microscope examination is essential. Whilst the sea is a suitable electrolyte, freshwater samples have been measured after the addition of 0.5% sodium chloride, which causes negligible changes in the cell volumes of the organisms. Applications of the instrument to marine studies have been described by SHELDON and PARSONS (1967), and for freshwater by EVANS and McGILL (1970).

6.8   Continuous recording

The Hardy Continuous Plankton Recorder works on a filtering principle. The flow of water over the apparatus rotates a stern propeller, and this rotation powers an internal mechanism whereby a long strip of bolting cloth is drawn across the path of water flowing through a tube in the apparatus. The planktonic organisms are trapped on the cloth, and the same driving mechanism winds the cloth onto a take-up spool in a formalin reservoir. The recorders are towed by ships making regular voyages, and a continuous record is thus obtained over considerable distances. Examination of the strip of bolting cloth in the spool, combined with information from the ship's log, will allow analysis of plankton distribution. These long hauls give information on the irregular distribution of plankton. The data built up over the years will also indicate any long-term changes in plankton abundance and composition. The recorder samples at one level in the sea; an apparatus of more recent design will move up and down whilst being towed, so sampling at different levels.

6.9   Biochemical methods

A number of biochemical analyses have been applied to measurements of phytoplankton biomass. These include estimates of total particulate carbon (non-specific for phytoplankton) by wet oxidation with chromic acid, or combustion of organic matter in oxygen followed by measurement of carbon dioxide yield. This method can be confused by the presence of inorganic carbonates. Colorimetric estimates of total phytoplankton carbohydrate have been applied, either by reaction with phenol and sulphuric acid (MARSHALL and ORR, 1962), or with anthrone (STRICKLAND and PARSONS 1968). The zooplankton contribution to total carbohydrate is small. It is not always possible to equate plant cell numbers with the quantities of carbohydrate measured, however. Measurements of DNA and ATP in particulate matter are used, and assay of plant enzymes has been proposed as a measure of phytoplankton.

6.10   The standing crop

The methods of counting described are aimed at obtaining quantitative data on the standing crop of plants. The standing crop at any one time is the balance resulting from the rate at which the plants have been growing and the rate of cropping. Cropping may result from grazing as in the sea, from outflow as with the larger diatoms in lakes, or loss of buoyancy and sinking. A measure of the standing crop will have a direct bearing on the productivity of a water sample (p. 83). As shown in earlier sections, changes in the standing crop of phytoplankton organisms are governed by a complex interplay of environmental conditions. Light and water turbidity determine both the variable depth of the photosynthetic zone and the light intensity distribution within it. The supply and recycling of nutrients (nitrate, phosphate and silicate and trace substances) will be rate limiting if any one falls below threshold values. Nutrient replenishment of the photosynthetic zone is dependent on turbulence, upwelling, the excretions of zooplankton and the activities of bacteria and fungi. Rate of cropping and degree of turbulence will be significant, particularly if large numbers of cells are carried below the photosynthetic zone for long periods. Temperature changes will lead to increased rates of growth in light and to higher respiratory losses in the dark. The cropping factor may also be altered in the sea by changes in the nature of zooplankton populations with temperature.

Any of the conditions outlined have an effect, sometimes major, on plant growth. Measurement of the standing crop at a particular time is a quantitative expression of the balance of factors then operating.

6.11   Measuring primary productivity

Plants, by their photosynthetic activity, are the primary producers of organic matter and the principal contributors to energy flow in ecosystems.

Measurements of primary production allow the dynamic aspects of an ecosystem to be elucidated. Comparative data from natural populations and cultivated plants will stimulate research into new sources of food and methods of obtaining more efficient production of organic matter.

6.11.1   Units of measurement

As already stated (p. 16), the new organic matter formed by a plant is its *gross production*. That part not used by the plant for its own respiratory needs is its *net production*. The *net production* of organic matter is available as for food for heterotrophs (animals, bacteria, fungi) either before or after death of the plants.

In general terms, the standing crop is the 'weight of plant material that can be sampled or harvested by normal methods, at any one time, from a given area' (WESTLAKE, 1963). The earlier definition of phytoplankton standing crop (p. 93) is similar. Measurement of phytoplankton biomass has to be made in terms of both the surface area of the water mass and the depth below this wherein photosynthesis can take place. This depth can be taken as the lower limit of the euphotic or photosynthetic zone (p. 20), the compensation depth (p. 23), or depth of the mixed layer which is often similar. If the mixed layer is much deeper than the photosynthetic zone then the total biomass will be underestimated if only that above the compensation depth is measured. Phytoplankton circulating in dark waters will also respire more organic matter than that synthesized. Hence with a very deep mixed layer there will be a marked reduction in *net production* by the phytoplankton community if the euphotic zone is shallow. This is seen in the low productivities of winter months.

Units of measurement (WESTLAKE, 1963)

Biomass—kilogrammes per square metre ($Kg\ m^{-2}$)
Annual production—metric tons (tonnes) per hectare ($mt\ ha^{-1}$)
Daily productivity—grammes per square metre ($g\ m^{-2}$)

6.11.2   Productivity measurements

Annual productivity is the most frequently used basis for comparison, either in terms of biomass per unit area (e.g. cultivated plants, terrestrial and aquatic communities of higher plants, and seaweeds), or as a measure of photosynthesis (e.g. oxygen produced or carbon dioxide used as with phytoplankton). Phytoplankton biomass measurement first requires separation of zooplankton. Dry-weight measurements made at 105 °C have been criticized on the basis of loss of some volatile components other than water and a temperature of 60 °C is preferred. Loss of weight of a dried sample after ignition at 550 °C gives the ash-free dry weight, the weight loss being regarded as oxidized organic matter. Calcareous substances will also decompose and this would introduce an error if coccolith-bearing organisms predominated. With diatoms, the ash (mainly SiO_2) constitutes some 50% of the dry weight, but for most other planktonic algae it is less than 5%.

6.11.3 Oxygen measurements

In photosynthesis the volume of oxygen produced nearly equals the volume of carbon dioxide fixed. Paired bottles, one transparent and the other darkened, are suspended at various depths in the sea or a lake, each containing raw water previously collected from the depth at which the bottle is placed. Alternatively laboratory incubations are carried out with temperature and illumination similar to those at the various depths from which samples were collected. Oxygen contents of the raw water samples are first determined (C_1). After suspension in the sea or lake for the required time the bottles are brought to the surface and the oxygen content again determined. Oxygen (C_2) in the dark bottles will be reduced because of the respiratory activity of enclosed organisms. Assuming that respiration rates are similar in the light and dark bottles, the difference in oxygen content between the light bottles (C_3) and the dark bottles is that due to photosynthesis.

$C_3 - C_2 =$ gross photosynthesis
$C_3 - C_1 =$ net photosynthesis
$C_1 - C_2 =$ respiratory activity (VOLLENWEIDER, 1969)

Oxygen can be measured quantitatively by the Winkler method, in which oxidation of a manganese salt causes the liberation of iodine which is then titrated with a standard sodium thiosulphate solution. This method has high accuracy with very small quantities of oxygen, and is widely used but also criticized. Lengthy incubations may harm enclosed phytoplankton, and bacterial growths are possible. Inaccuracies may also arise due to manipulation errors.

6.11.4 Carbon dioxide uptake using ^{14}C

STEEMANN NIELSEN (1952) first used ^{14}C in a productivity study of oceanic waters. The isotope is a weak β emitter with a very long half-life (ca. 4700 years). Light and dark bottle incubations, either in sea or lake or on board ship with a suitable light source, are carried out with raw water samples from the depths required. Samples are stored in a dark box and their carbon dioxide content determined. The labelled carbon is added $(NaH^{14}CO_3, Na_2{}^{14}CO_3)$ and the light and dark bottles suspended at the various depths, or incubated in lighted containers on board ship. At the end of the incubation period the plankton is isolated on a membrane filter and then placed on counting planchettes and stored in a desiccator. The activity of the filtered sample is then determined with a suitable measuring instrument (VOLLENWEIDER 1969). The dark bottle is used to check that there is no significant dark fixation of ^{14}C.

Criticisms have also been levelled at this method. The artificial light or shaded daylight used in shipboard determinations is regarded as atypical compared with underwater illumination. $^{14}CO_2$ is not assimilated at the same rate as $^{12}CO_2$ and a correction factor is required. Dark fixation may not be fully corrected for in dark-bottle measurements. Release of fixed carbon as extracellular products (p. 78) is another likely source of error. Whilst this may be less than 1% of fixed carbon in eutrophic lakes, as much as 35% may be released in oligotrophic waters (FOGG, NALEWAJKO and WATT, 1965). Photoinhibition will release large quantities. Phytoplankton photosynthesis shows diurnal variations which will influence short incubation periods. What is measured has been called gross production, net production or something between the two.

Despite the necessary corrections, the ^{14}C method is more sensitive than oxygen measurements for waters of low productivity. It has facilitated productivity measurements in many natural waters. It would seem essential that critical evaluations should be made with both cultures and field populations.

6.12   Phytoplankton productivity—some results

Phytoplankton populations stand low in productivity compared with natural vegetation and cultivated plants (PHILLIPSON, *Ecological Energetics,* p. 46). Primary production of deep oceans (less than 1.0 gC/m²/day) is about twice that of deserts. In shallow seas (0.5–3.0 gC/m²/day) there is a productivity similar to deep lakes. In eutrophic lakes productivities of 3–10 gC/m²/day are recorded—approaching that of forests, grassland and cultivated plants. Annual productivities of natural phytoplankton in no way approach those of crop plants. *Scenedesmus quadricauda* mass cultured in sewage treatment works shows productivity resembling that of cultivated plants (45 mt ha^{-1}). Whilst phytoplankton productivity is low compared with agricultural crops, the costings of labour, fertilisers and capital investment must be considered with the latter. With 70% of the earth's surface covered by sea, marine phytoplankton clearly make a major contribution to overall productivity.

6.13   Mathematical models

Attempts have been made to assess the significance of the changing parameters affecting phytoplankton standing crops and productivity. Various mathematical expressions have been formulated for the rates of change of populations from measurements of temperature, transparency, input of radiant energy, depth of mixed layer and quantities of zooplankton, (RILEY and BUMPUS, 1946; RILEY, STOMMEL and BUMPUS, 1949; STEELE, 1956). These mathematical models appear applicable to predictions of primary productivity, but doubts have been expressed regarding their value in terms of productivity at higher levels in the food chain. Use of mathematical models in the initial stages of surveys as a means of finding the more fruitful lines of investigation would seem worthwhile.

7 Man-made Effects

7.1 Eutrophication

Eutrophication is the enrichment of a water mass with inorganic and organic plant nutrients. Whilst often a feature of lakes and freshwater impoundments, it also occurs in some estuaries and coastal waters. Natural eutrophication is a slow process of enrichment and is part of an ageing process. Artificial eutrophication follows discharge of domestic and industrial effluent, and the run-off from agricultural land dressed with nitrogenous artificial fertilisers. In nature eutrophication is a beneficial process in that it produces enhanced productivity in the water mass. The accelerated enrichments of water masses by artificial means followed by 'population explosions' (blooms) of planktonic organisms and filamentous algae, have serious side effects. Death and decay of the surface 'blanket' of algae will cause oxygen depletion with death of animals. The odours associated with decay will taint the water and make the surroundings unpleasant. Where 'blooms' are due to toxin-producing blue-green algae there is an additional hazard to animals. The amenity value of lakes and impoundments will be reduced and the technical difficulties in water purification for drinking or industrial usage will be considerable. The Zurichsee in Switzerland has been subject to seasonal blooms of the blue-green alga *Oscillatoria rubescens* following increased sewage discharge due to new building developments on the shores. Blue-green algae cause blockage of sand filters. The accompanying bacterial populations will require that the water be chlorinated. Artificial eutrophication is more a feature of impounded waters, although large algal populations can develop in slow flowing rivers and shallow estuaries.

The causative factors of this explosive algal growth are usually attributed to nitrate and phosphate. For some Wisconsin lakes it has been stated that nuisance 'blooms' of phytoplankton algae can be expected if the phosphate phosphorus concentrations exceed 0.01 mg dm^{-3}, and the nitrate nitrogen concentration is more than 0.3 mg dm^{-3}. Regular monitoring might then allow some form of 'early warning' system. The blue-green alga *Oscillatoria rubescens* has been proposed as a 'biological indicator' of rapid eutrophication, but it seems unlikely that it will be an effective agent for such predictions. Reduction of the nitrate and phosphate 'loads' has been regarded as a high priority, but the methods are costly. Detergents are a source of a good deal of the phosphate discharged in domestic effluents. Up to 50% of the phosphate content in some effluents in traceable to this source. Nitrate is the principal causative agent in artificial fertilisers. Nitrate dressings of land are likely to increase in the future, so that some attention will have to be given to side effects on standing freshwater. Some indications of the long-term subtle effects of eutrophication are already apparent. Phytoplankton periodicity in Lake Erie in 1935 showed a typical annual cycle with peaks in spring and autumn. In 1962, following increased eutrophication, the spring and autumn peaks became merged with a general increase in algal cell numbers in both winter and summer.

A study of eutrophication in a number of small ponds has produced interesting data on pH changes associated with dense growths of algae (O'BRIEN and NOYELLES, 1973). Observations were also made on 'control' ponds of similar capacity. Growths of mainly green algae (*Scenedesmus quadricauda, Oocystis parva, Tetraspora lacustris* and *Pandorina morum*) during the summer months induced changes in pH. The zooplankton organism *Ceriodaphnia reticulata* disappeared from the pond water at high pH values—irrespective of the numbers of algae present. The animals appeared susceptible to low thresholds of pH change. A change from pH 10.8 to 11.2 was critical—all animals being killed within 18 h after initial subjection to water of the higher pH value. These changes occur in waters of low buffer capacity, but the results suggest that pH monitoring may be an effective method of predicting side effects of algal blooms on certain zooplankton.

7.2 Man-made lakes

Man-made lakes are large masses of standing water (usually fresh), created by impounding the run-off from a catchment area. Often this entails building a dam at the end of a river valley. The lake then forms fairly rapidly, inundating large tracts of land. Often the filling stage is preceded by burning the vegetation. Both this and the inundated soil contribute a high initial fertility to the water mass when cultivated land is flooded. This initial eutrophy may be reflected in abundant growths of phytoplankton. Flooding an acid moorland valley, however, is followed by little noticeable phytoplankton growth. In fertile lakes the initial instability of the phytoplankton may be seen in the changing populations. Another result of the plant 'population explosion' may be seen in formation of 'sudd'—a floating mass of living and dead vegetation. Over-production of plants and their subsequent death and decay will bring about deoxygenation. Man-made lakes have been created as storage reservoirs for drinking water, to control river flow and help in irrigation, and to supply a sufficient head of water for electricity generation. In many localities they provide a recreational amenity. On a very large lake (e.g. Lake Kariba) a fishing industry may be developed. The advantages of such overall schemes to underdeveloped countries can be appreciated. It takes many years fully to evaluate the biological changes involved in such ventures and the overall implications for the area. Continued monitoring of the biological, physical and chemical data of newly-established lakes will give information of great interest to fishery biologists. The data may also help in the planning of similar ventures. The phytoplankton are the most significant biological indicators. All too often, however, insufficient attention is given to this aspect of lake establishment, and valuable information is lost.

7.3 Potable water supplies

With the increasing demands of both the domestic consumer and industry, the supply of fresh water is likely to become a major problem. The supply of drinkable water will clearly be very important. Water of the purity required for human consumption is also needed for delicate processes in many light industries (e.g. photography, paper-making and soft-drink manufacture). For heavy industries the water need not be so pure. To maintain a 'constant head' of water supply many authorities rely on storage reservoirs or impoundments. In some cases a natural lake is a city's water supply (e.g. Loch Katrine for Glasgow). Water drainage from agricultural land is rich in plant nutrients, and rivers which receive domestic effluents, whilst unsuitable as a direct source of drinking water, are potentially rich media for algal growth. Once these enriched waters are impounded, thermal stratification in the summer may result in 'blooms' of algae in the epilimnion, accompanied by deterioration in the condition of the hypolimnion. To the water engineer the problems are not only those of water colour, taste and smell, and the effects of 'blooms', but also the levels in the reservoir from which the water should be drawn off. At the level of the epilimnion during periods of algal bloom there will be filtration difficulties, as well as an increased bacterial 'load' due to death and decay of algal cells. The hypolimnion similarly may be unpleasant because of its high organic content, together with sulphides and metallic ions. Whilst it would seem best to site the outlet in the region of the metalimnion, this is a variable feature of lake stratification, and additionally the demands on the reservoir are greatest at times of maximum thermal stratification in summer and when fresh supplies from catchment areas are at a minimum. The oscillation of water masses accompanying seiche formation (p. 26) also create difficulties because epilimnion water with high algal content and hypolimnion water with organic matter and metal compounds may alternate at the outlet. Hence, in designing reservoirs and siting pipes for withdrawing water, account has to be taken of the likely internal movements of the water impounded. Sufficient information is available today for fairly accurate estimates of the levels of seiche formation and oscillation. Regular biological and chemical monitoring of the water is a necessary feature of its management. The problems of artificial eutrophication are applicable to all forms if impoundments, and nutrient-enriched waters from a catchment area will be a potential source of algal blooms troublesome for water engineers.

7.4 Fishponds

These are relatively small water masses which are centres of intensive food production. In some parts of the world (Asia, Africa and Eastern Europe) they are important sources of protein. The food webs are shorter and less complicated than those in lakes. The high productivity of fishponds calls for intensive 'cultivation', with application of large quantities of fertiliser. In poor countries animal and human excreta are utilized; in richer areas artificial fertilisers are applied. Management of a fishpond calls for a marked degree of eutrophication, and in eastern Europe a dense surface bloom of blue-green algae (particularly one dominated by *Aphanizomenon flos-aquae*) is regarded as a satisfactory indicator of potential productivity. Such blooms can be produced by application of phosphates. The odd feature of these blooms is that blue-green algae *en masse* are known to produce toxic substances (p. 79) and may not be the most suitable food for animals. Fish mortalities have been recorded following the disappearance of blue-green algae blooms, but this may have resulted from deoxygenation or production of harmful decomposition products. Possibly the continuous bloom of blue-green algae represents a steady-state condition in which the flow of food continues via their remains, with associated bacteria and zooplankton organisms. In the Far East, the milkfish *Chanos chanos* is grown in brackish fishponds on alluvial tidal flats rich in plant nutrients. The abundant growths of algae encouraged by such conditions serve as food for the milkfish. Experiments on the digestion of the various algae by fish showed that planktonic phytoflagellates were most nutritious, followed by the diatoms and blue-green algae. Filamentous green algae were less readily digested. It is uncertain whether milkfish are plankton feeders. In experiments where large masses of phytoflagellates were supplied as the sole food source there was little evidence from analysis of stomach contents of direct utilization. Algae growing or settled on the pond floors are probably the best food source for the sucking action typical of the fish's feeding habit. A surface 'scum' of algae can be engulfed in a similar way. In some cases the fishpond phytoplankton is important as food for the young (fry) stages (e.g. as with *Tilapia*—which it reared in many tropical and subtropical countries). Many of the fish cultivated do not directly utilize phytoplankton as the food source. However, the silver carp shows an extremely rapid growth rate in freshwater fishponds, and is exclusively a phytoplankton feeder. Hence there is some interest in developing systems of fishpond cultivation using this organism. The food chain is then effectively shortened.

7.5 Pesticides

It has been said that wholesale use of pesticides of the DDT type is one of man's most disastrous large-scale experiments. Evidence cited includes the reported accumulation of DDT and related compounds in the body fat of penguins and seals in Antarctica—many thousands of miles from centres of human activity. Evidence that pesticides increase in concentration at successive steps in food webs is frequently quoted. With the phytoplankton occupying such a key position in food webs there is much speculation on the accumulation of toxic pesticides at this level. Precise information is limited, however. Of greater significance is the observation that the photosynthesis of marine phytoplankton is affected by very low concentrations of DDT (WURSTER, 1968). Cultures of phytoplankton organisms with cell numbers simulating natural populations showed significant reductions in ^{14}C fixation in the presence of concentrations of DDT as low as 1–2 ppb. Similar results were obtained with a natural population under identical experimental conditions. The range of DDT concentrations used (1–100 ppb) was reported as being similar to that found in some inshore waters. These experiments highlight the sub-lethal effects of certain pollutants. Reduction in photosynthesis is a toxic stress effect which could ultimately alter the species balance of a phytoplankton community. This in turn could lead to the selective proliferation of certain species, and trigger off periodic algal blooms accompanying eutrophication of the water mass. Changes in the qualitative nature of the phytoplankton could bring about alterations in the balance of the primary and secondary consumers.

It can be argued that the results of laboratory experiments need to be interpreted with caution. Conditions in small flasks are artificial and can never really simulate those in a natural environment. However, the clear pointer from these experiments is that a sub-lethal effect of pesticide pollution could be a reduction of the photosynthetic efficiency of phytoplankton. The consequences could be significant both at the nutritional level and in terms of the oxygen produced as a by-product of photosynthesis. The contribution of the phytoplankton is here of major significance. The problem has probably eased in recent years because of the decreased usage of DDT. Concentrations of the order stated above do not exist in the open sea. However, the warnings implicit in such an experiment should not be ignored.

7.6　Oil spills and emulsifiers

Large numbers of tankers transport crude oil from areas of production to refineries thousands of miles away. With a steady rise in demands for petroleum fuels, tanker sizes have also increased. Loss of crude oil from tankers through wrecks, collisions and accidental spillages now constitutes a major feature of sea pollution. Crude oils are complex mixtures, variable in their composition. When oil spreads on water certain fractions evaporate. It has been estimated that of the crude oil lost from the Torrey Canyon after its wreck on the Seven Stones rocks, 15 miles west of Land's End, about a third evaporated from the sea surface. Loss of material by evaporation results in a stable water-in-oil emulsion ('mousse') which floats on the surface. Oil spills at sea appear to have little effect on the plankton. Zooplankton are less harmed than phytoplankton. Certain diatoms (*Ditylum brightwellii, Coscinodiscus granii, Chaetoceros curvisetus*) were found to be sensitive to 100 ppm of kerosene and fuel oil. Other species (*Grammatophora marina, Melosira moniliformis*) were found to tolerate concentrations of 1%. The toxic components of the crude oils include water-soluble phenols and volatile aromatic hydrocarbons; these constitute a major part of the crude oil lost by evaporation.

Oil-spill emulsifiers convert the viscous water-in-oil emulsion into a milk-like oil-in-water emulsion. Experience with the Torrey Canyon oil spill showed that the emulsifiers used were highly toxic to many intertidal plants and animals. Away from coastal regions some injurious effects on phytoplankton were observed, noticeably with green flagellates of the Prasinophyceae, and with some diatoms and dinoflagellates. Some colourless flagellates (including members of the Cryptophyceae) appear able to grow in sea water containing quantities of toxic emulsifiers harmful to other organisms when cultured under laboratory conditions. A number of emulsifiers (also called dispersants) have now been manufactured which are far less toxic than those used at the time of the Torrey Canyon disaster. Toxicity tests with such compounds should always include representative organisms of the phytoplankton. A dispersant, even if apparently non-toxic to larger organisms, may prove lethal to nanoplankton flagellates.

7.7   Detergents

Side effects of phosphate-containing detergents in domestic sewage include foam formation and eutrophication. Many detergents contain surface-active compounds. Some of these are not broken down by bacterial action. Detergents are of three main types—non-ionic, anionic and cationic. A number of detergents have been subjected to laboratory tests with unicellular marine green algae (UKELES 1965). A variable susceptibility was observed with those non-ionic, with significant growth inhibitions at concentrations between 10 and 100 ppm. The anionic forms—the most widely used—showed significant growth inhibitions in the concentration range 10–100 ppm. With the cationic detergent lauryl pyridinium chloride the growth of all algae tested was completely inhibited at 100 ppm, and with many species a significant inhibition occurred at 10 ppm. With flagellates lacking cell walls a 50 per cent growth inhibition was obtained in 0.01 ppm of this detergent. As would be expected, toxic effects were greater with increased surface activities. Algae with thicker cell walls tended to be the more tolerant although this would also depend on the chemical nature of the cell walls. 'Golden-brown' flagellates of the Chrysophyceae and Haptophyceae were more susceptible than the green algae.

These experiments have also been concerned with sub-lethal effects. Damage to natural populations in the long term will follow periodic exposures to sub-lethal toxic concentrations. The difficulties lie in equating laboratory experiments with conditions in nature. Whilst detergent concentrations in the ranges 0.1–0.6 ppm and 3–6 ppm have been detected in rivers we have no information on the dilution effect in the sea or lakes. Nevertheless we can take heed of the experimental data quoted. It does indicate that certain planktonic algae are at risk and some selective species disappearances could occur.

7.8   Heavy metals and radioactive substances

7.8.1   Heavy metals

Copper compounds have long been used as algicides to destroy algal growths in ornamental ponds, and have been incorporated into anti-fouling paints for ships hulls. Whilst acting as a protection against the attachment of algae or barnacles, the quantities of toxic metal leaching into the water are very small and unlikely to affect phytoplankton when the ship is moving. Cadmium, mercury and lead released into rivers and estuaries could prove more harmful. Numerous man-made sources of mercury are known (fungicides in agriculture and in pulp and paper mills; insecticides, vermicides, rodenticides; by-products of industrial processes). Concentrations of mercury of 1 ppm have been found to inhibit algal growth and metabolism. Organic mercury compounds are severely toxic at much lower concentrations. Mercury in the sea is converted to methyl mercury $((CH_3)_2Hg)$—a compound strongly lipophilic. Enclosed shallow sea areas (Japan and Sweden) are known to be more at risk, particularly with the mercury becoming concentrated at successive stages in the food chain. Oligotrophic lakes are more affected by mercury pollution than those which are eutrophic. Lead pollution from mining operations has in the past contaminated some rivers to the extent that they were devoid of plant life and animals. Removal of lead pollution has brought life back to some rivers (e.g. the Ystwyth and Rheidol in mid-Wales). Heavy metal pollution of sea, rivers and lakes will need to be watched carefully. Sub-lethal effects on phytoplankton may have long-term results which could be serious.

7.8.2   Radioactive substances

The sea is the main recipient of radioactive waste from weapon testing fall-out, isotope manufacture and nuclear power generation. Scrupulous care is certainly used in disposal, but this is the pollutant source offering the gravest risk to man. There is clear evidence that most of the radioactive material is adsorbed on suspended inorganic particles (Table 8). The phytoplankton will also concentrate radioactive material, and migrating zooplankton will feed at night on phytoplankton which have absorbed radioactive material by day, voiding radioactive faecal pellets, which, with contaminated plant and animal remains, will become incorporated into bottom deposits in the sea and then undergo long term recycling from this source.

Table 8   Average surface areas of particulate matter per cubic metre of sea water. (Data from ZENKEVITCH (1960), Disposal of Radioactive Waste Conf. Proc., Monaco, 100–101).

Inanimate particles	25 m²
Bacteria	0.05 m²
Phytoplankton	0.025 m²
Zooplankton	0.0002 m²

7.9   Warm-water effluents

Use of waters from rivers and estuaries for cooling purposes in electricity-generating plants has raised questions regarding the ecological effects of heated water in the environment. These include de-oxygenation and the effects of warm water on organisms of limited temperature range. In rivers and lakes and shallow estuaries these effects could prove serious. Long-term raising of the temperature by a few degrees may have long term subtle effects, leading to the establishment of organisms from warmer waters.

7.10   Man and phytoplankton

It is fitting to close this general survey of the phytoplankton by referring to attempts by man either to use directly or to improve the yields of the primary producers. Direct usage of plankton by man usually applies to zooplankton. Thor Heyerdahl in his account of the Kon-Tiki expedition (1950) described the netting of plankton where the cold Humboldt current turned west south of the equator. Several pounds of 'plankton porridge' were collected every hour. The appearance and taste depended on the dominant organisms, with tastes described as simulating shrimp paste, lobster or crab, or even caviare. Noticeably, however, the 'uneatable' plant component was too small to be collected by the nets used. This factor of edibility has to be reckoned with. Whilst food values in terms of protein, carbohydrate and fat are high, in many cases the algae presented a taste barrier too difficult to overcome. However, it has often been suggested that unicellular algae grown in mass cultures could be harvested and the dried products incorporated into other foods where the taste factor would be masked, e.g. incorporation of the dried algae in animal feeding stuffs. Mass cultures of *Chlorella pyrenoidosa* have been grown using solar radiation in large polythene tubes, with nutrient media 6.325 cm deep supplied with air containing 5% carbon dioxide. The shallow layer of culture medium ensured adequate illumination of the algal cells. The liquid was kept in motion by pumps, and temperatures stabilized by heat interchangers. In test runs a plant-tissue yield equivalent to 12.5 tonnes dry weight per ha was obtained. With better radiation input a trebling of the yield has been envisaged. Even if only 50% of this is achieved it would represent almost an eightfold increase over that obtained with cultivated land. The problem of subsequent usage—direct or indirect—cannot be ignored, however. It may also be questioned whether *Chlorella* is the most suitable organism to use. Work with invertebrate larvae has shown that with this alga problems of digestion—probably due to the nature of the cell wall—lowers its value as food (p. 73). Green flagellates lacking cell walls, which are good food sources for invertebrates, might prove more digestible by vertebrates. Alternatively, any routing of algal nutriment to man *via* farm animals may be unnecessarily wasteful. Bivalve molluscs have a high growth efficiency and can be cultivated on a large scale using mass cultures of algae. A more efficient use of algal protein would then be obtained.

Perhaps the ultimate design for algal mass culture is that envisaged for space travel. The cultures of green algae would form a closed system with the space capsule, the nitrogenous waste of the travellers being diluted and circulated as algal nutrients, and the respired air used as a carbon dioxide source. In exchange the travellers could utilize the oxygen released by photosynthesis and benefit from the food value of the cultivated algae. The problems of plant design and illumination are considerable, however.

In recent years increasing attention has been given to marine fish farming. An inevitable problem with marine pisciculture is how the area under cultivation can be effectively fertilized without loss of plant nutrients by lateral transport. Fertilization of Loch Craiglin in the west of Scotland with sodium nitrate and superphosphate at various times of the year produced dramatic increases in the phytoplankton standing crop. Increases in zooplankton were observed and a threefold increase in benthic populations. In Kyle Scotnish similar enrichments produced increases in phytoplankton, zooplankton and fivefold increases in growth rates were obtained with introduced plaice. Fish farming in association with warmed water effluent from power stations is under investigation. A similar use of natural sea 'basins' has been proposed in the 'coral corral' method of cultivation. Nutrient-rich deep water would be pumped into circular atoll lagoons for the cultivation of phytoplankton and zooplankton. Captive baleen whales, which have a high efficiency for conversion of zooplankton to protein, would then form the ultimate stage in the food chain (PINCHOT, 1966). Nutrient-rich water from regions of upwelling could be pumped into ponds (on land) of large surface area. Solar radiation would be used for phytoplankton cultivation. Feasibility studies have suggested that up to 100 tonnes dry weight of plankton/hectare/annum could be produced (ROELS, GERARD and BÉ, 1971) by using the nutrients present in the deep water without additional enrichments.

Whatever the future developments may be in fish cultivation, we come full circle with the inevitable point that marine phytoplankton represents one of the world's major organic resources. Inshore exploitation and cultivation will have but fringe effects on the great bulk of drifting plant life. Conservation of this natural resource should be our constant care; we may never know when man's interference has passed a point of no return.

Appendix: Culture Media

Numerous culture media have been devised either for general usage or specifically for certain algae. The following media will support the growth of planktonic algae.

(1) Marine–Erdschreiber medium

Seawater	$1 \, dm^3$ (It is an advantage to allow the seawater to 'age' for a few months before use)
Soil extract	$50 \, cm^3$
Sodium nitrate	$NaNO_3$ 0.2 g
Disodium hydrogren phosphate	$Na_2HPO_4.\ 12H_2O$ 0.03 g

Filter the seawater through Whatman No. 1 paper. Pasteurize by heating to 75 °C, keep at this temperature for a short while, then leave to cool. This treatment is repeated. It is an advantage to add $100 \, cm^3$ of distilled water to $900 \, cm^3$ of seawater to prevent formation of precipitates.

Soil extract is prepared by air drying a sample of garden soil. The dry soil is then crushed and passed through a fine sieve to remove stones. The sieved material is autoclaved with twice its volume of supernatant water, the autoclaved mixture left for several days for the soil to sediment, and the yellow-brown coloured supernatant then decanted. This must be stored in a flask plugged with non-absorptive cotton wool.

In preparing the medium, the solutions of sodium nitrate, sodium phosphate and soil extract should be separately autoclaved before mixing with the pasteurized seawater. If the sodium phosphate is added to the seawater before pasteurization it may cause formation of a precipitate.

(2) Freshwater. Modified Chu No. 10 (Gerloff)

Calcium nitrate	$Ca(NO_3)_2$	0.4 g
Dipotassium hydrogen phosphate	K_2HPO_4	0.1 g
Sodium carbonate	Na_2CO_3	0.2 g
Magnesium sulphate	$Mg \, SO_4.7H_2O$	0.25 g
Sodium silicate	Na_2SiO_3	0.25 g
Ammonium ferric citrate		0.05 g
Distilled water—$1 \, dm^3$		

Further sources of information on culture media:

Culture Collection of Algae and Protozoa—List of Strains (1971). 36 Storey's Way, Cambridge, CB3 0DT.

PRINGSHEIM, E. (1948). *Pure Cultures of Algae—Their Culture and Maintenance.* University Press, Cambridge. A small book of fundamental importance which should be read by anyone contemplating culture studies with algae.

PROVASOLI, L., MCLAUGHLIN, J. J. A. and DROOP, M. R. (1957). The development of artificial media for marine algae, *Arch. Mikrobiol.,* **25,** 392–428. A detailed account of numerous media devised for marine algae prior to date of publication. An essential reference for culture studies with algae.

STEIN, J. (Ed.) (1973) *Handbook of phycological methods: Culture methods and growth measurements.* University Press, Cambridge.

References

BAINBRIDGE, R. (1953). *J. mar. biol. Ass.*, **32**, 385–445.
BALLANTINE, D. (1953). *J. mar. biol. Ass.*, **32**, 129–147.
BIBBY, B. T. and DODGE, J. D. (1972). *Br. phycol. J.*, **7**, 85–100.
BROOK, A. J. (1964). Phytoplankton, in *The Vegetation of Scotland* (ed. Burnett, J. H.), 290–305. Oliver & Boyd, Edinburgh and London.

CANTER, H. and LUND, J. W. G. (1948). *New Phytol.*, **47**, 238–261.
CANTER, H. and LUND, J. W. G. (1951). *Ann. Bot.*, **15**, 359–371.
CANTER, H. and LUND, J. W. G. (1966). *Verh. int. Verein. Limnol.*, **16**, 163–172.
CANTER, H. and LUND, J. W. G. (1968). *Proc. Linn. Soc. Lond.*, **179**, 203–219.
CANTER, H. and LUND, J. W. G. (1969). *Öst. bot. Z.*, **116**, 351–377.
CARLUCCI, A. F. and BOWES, P. M. (1970a). *J. Phycology*, **6**, 351–357.
CARLUCCI, A. F. and BOWES, P. M. (1970b). *J. Phycology*, **6**, 393–400.
CHRISTENSEN, T. (1966). *Alger.* In *Botanik Bol. 2* (Systematiok Botanic). 180 pp. Munksgaard, Copenhagen.
CUSHING, D. H. (1964). In *Grazing in Terrestrial and Marine Environments* (ed. Crisp, D. J.), 207–225. Blackwell Scientific Publications, Oxford and Edinburgh.

DROOP, M. R. (1967). *Br. phycol. Bull.*, **3**, 295–297.

EVANS, J. H. and McGILL, S. M. (1970). *Hydrobiologia*, **35**, 401–419.

FOGG, G. E. (1963). *Br. phycol. Bull.*, **2**, 195–205.
FOGG, G. E., NALEWAJKO, C. and WATT, W. D. (1965). *Proc. R. Soc. B.*, **162**, 517–534.

GORHAM, E. (1964). In *Algae and Man,* (ed. Jackson, D. F.), 307–326. Plenum Press, New York.
GROSS, F. (1937). *Phil. Trans. B,* **228**, 1–47.
GROSS, F. and ZEUTHEN, E. (1948). *Proc. R. Soc. B.*, **135B**, 382–389.

HARVEY, H. W. (1934). *J. mar. biol. Ass.*, **19**, 761–773.
HARVEY, H. W. (1963). *The Chemistry and Fertility of Sea Waters,* 240 pp. University Press: Cambridge.
HOLMES, R. W. (1962). *US Fish Wildlife Serv.: Spec. Sci. Rep. Fish.* No. 433, 6 pp.
HUTCHINSON, G. E. (1944). *Ecology,* **25**, 3–26.
HUTCHINSON, G. E. (1967). *A Treatise on Limnology,* Vol. 2, xi + 1115 pp., John Wiley & Sons, New York & London.

JERLOV, N. G. (1951). *Rept. Swedish Deep Sea Expedition*, **3**, 1–59.

JOHNSTON, R. (1963*a*). *J. mar. biol. Ass.*, **43**, 409–426.

JOHNSTON, R. (1963*b*). *J. mar. biol. Ass.*, **43**, 427–456.

KNIGHT-JONES, E. W. (1951). *J. Cons. perm. int. Explor. Mer.*, **17**, 140.

LEEDALE, G. F. (1967). *Euglenoid Flagellates*. Prentice-Hall Inc., Englewood Cliffs, N.J.

LEFEVRE, M., JACOB, H. and NISBET, M. (1952). *Annls St. Cent. Hydrobiol. Appl.*, **4**, 5–197.

LOVEGROVE, T. (1961). *J. Cons. perm. int. Explor. Mer*, **25**, 279–284.

LUCAS, C. E. (1947). *Biol. Rev.*, **22**, 270–295.

LUND, J. W. G. (1951). *Hydrobiologia*, **3**, 390–394.

LUND, J. W. G. (1954). *J. Ecol.*, **42**, 151–179.

LUND, J. W. G. (1959). *Br. phycol. Bull.*, **1 (7)**, 1–15.

LUND, J. W. G. (1964). *Verh. int. Verein. Limnol.*, **15**, 37–56.

LUND, J. W. G. (1965). *Biol. Rev.*, **40**, 231–293.

LUND, J. W. G., KIPLING, C. and LE CREN, E. D. (1958). *Hydrobiologia*, **11**, 143–170.

MARGALEF, R. (1958). In *Perspectives in Marine Biology* (ed. Buzzati-Traverso, A.A.), 323–349. University of California Press, Berkley and Los Angeles.

MARSHALL, S. M. and ORR, A. P. (1962). *J. mar. biol. Ass.*, **42**, 511–519.

MENZEL, D. W. and SPAETH, J. P. (1962). *Limnol, Oceanogr.*, **7**, 151–154.

MOORE, J. K. (1963). *Limnol. Oceanogr.*, **8**, 304–305.

MUNK, W. H. and RILEY, G. A. (1952). *J. mar. Res.*, **11**, 215–240.

NEWTON, L. (1959). *Trans. Proc. bot. Soc. Edinb.*, **38**, 141–150.

NYGAARD, G. (1949). *K. danske. Vidensk. Selsk. Biol. Skr.*, **7**, 293 pp.

O'BRIEN, W. J. and NOYELLES, F. (1973). *Ecology*, **53**, 605–614.

PARKE, M. and MANTON, I. (1967). *J. Mar. Biol. Assoc.*, **47**, 445–464.

PEARSALL, W. H. (1932). *J. Ecol.*, **20**, 241–262.

PINCHOT, G. B. (1966). *Pt. Perspec. Biol. Med.*, **10**, 33–43.

POOLE, H. H. and ATKINS, W. R. G. (1937). *Proc. Roy. Soc. B*, **123**, 151–165.

PROCTOR, V. W. (1957). *Limnol. Oceanogr.*, **2**, 125–139.

PROVASOLI, L. (1971). In *Fertility of the Sea* (ed. Costlow, J. D.), 369–382. Gordon and Breach, New York.

RILEY, G. A. and BUMPUS, D. F. (1946). *J. mar. Res.*, **6**, 33–47.

RILEY, G. A., STOMMEL H. and BUMPUS, D. F. (1949). *Bull. Bingham oceanogr. Coll.*, **12**, 1–169.

ROELS, O. A., GERARD R. D. and BÉ, A. W. H. (1971). In *Fertility of the Sea* (ed. Costlow, J. D.), 401–416. Gordon and Breach, New York.

ROUND, F. E. (1971). *Mitt. int. Verein. theor. angew. Limnol.*, **19**, 70–99.

SHELDON, R. W. and PARSONS, T. R. (1967). *A Practical Manual on the Use of the Coulter Counter in Marine Science.* 66 pp. Coulter Electronics: Canada.

SMAYDA, T. J. (1970). In *Oceanography and Marine Biology: An Annual Review* (ed. Barnes, H.), No. 8, 353–414.

STEELE, J. H. (1956). *J. mar. biol. Ass.*, **35**, 1–33.

STEELE, J. H. and YENTSCH, C. S. (1960). *J. mar. biol. Ass.*, **39**, 217–226.

STEEMANN NIELSEN, E. (1952). *J. Cons. perm. int. Explor. Mer*, **18**, 117–140.

STRICKLAND, J. D. H. (1972). In *Oceanography and Marine Biology: An Annual Review* (ed. Barnes, H.), No. 10, 349–414.

STRICKLAND, J. D. H. and PARSONS, T. R. (1968). *Bull. Fish. Res. Bd Canada*, No. 167, 311 pp.

SVERDRUP, H. U., JOHNSON, M. W. and FLEMING, R. H. (1946). *The Oceans: Their Physics, Chemistry and General Biology*, 1087 pp. Prentice-Hall, New York.

TALLING, J. F. (1969). In *A Manual on Methods for Measuring Primary Production in Aquatic Environments* (ed. Vollenweider, R. A.), 22–25. Blackwell Scientific Publications, Oxford and Edinburgh.

TALLING, J. F. and DRIVER, D. (1963). *Proc. Conf. Primary Productivity Measurement, Marine and Freshwater*, 142–146.

UKELES, R. (1965), *J. Phycology*, **1**, 102–110.

UTERMÖHL, H. (1931). *Verh. int. Verein. Limnol.*, **5**, 567–596.

UTERMÖHL, H. (1958). *Mitt. int. Verein. theor. angew. Limnol.*, **9**, 1–38.

WESTLAKE, D. F. (1963). *Biol Rev.*, **38**, 385–425.

WOOD, E. F. J. (1955). *J. Cons. perm. int. Explor. Mer*, **21**, 6–7.

WOOD, E. F. J. (1962). *Limnol. Oceanogr.*, **7**, 32, 35.

WURSTER, C. F. (1968). *Science*, **158**, 1474–1475.

Further Reading

Specialized

FOGG, G. E. (1965). *Algal cultures and phytoplankton ecology.* Athlone Press: London.

HARVEY, H. W. (1963). *The Chemistry and Fertility of Sea Waters.* University Press: Cambridge.

HILL, M. N. (ed.) (1963). *The Sea,* Vol. 2, Chapters 7, 8, 17, 20. Allen & Unwin: London.

HUTCHINSON, G. E. (1967). *A Treatise on Limnology,* Vol. 2, Chapters 18, 20, 21, 22, 23. John Wiley: New York & London.

MACAN, T. T. (1970). *Biological Studies of the English Lakes,* Chapters 4, 5, 6, 13. Longmans: London.

ODUM, E. (1959). *Fundamentals of Ecology* (2nd edition). Saunders: Philadelphia.

RAYMONT, J. E. G. (1963). *Plankton and Productivity of Oceans,* Chapter 1–11. Pergamon: Oxford.

SVERDRUP, H. U., JOHNSON, M. W. and FLEMING, R. H. (1946). *The Oceans, their Physics, Chemistry and Biology,* Chapters 16, 19. Prentice-Hall: New York.

General

PHILLIPSON, J. (1966). *Ecological Energetics.* Edward Arnold: London.

ROUND, F. E. (1973). *The Biology of the Algae,* Chapters 4, 5, 6, 7, 8, 9, 12. Edward Arnold: London.

TAIT, R. V. (1968). *Elements of Marine Ecology,* Chapters 1, 2, 3, 4, 5, 8. Butterworths: London.

WIMPENNY, R. S. (1966). *The Plankton of the Sea,* Chapters 5, 6, 7, 8, 9, 11, 12. Faber: London.

WOOD, E. F. J. (1965). *Marine Microbial Ecology.* Chapman & Hall: London.

Important Review Articles

LUND, J. W. G. (1965). The ecology of the freshwater phytoplankton, *Biological Review,* **40,** 231–293.

STRICKLAND, J. D. H. (1972). The marine planktonic food-web. In *Oceanography and Marine Biology: An Annual Review* (ed. Barnes, H.). No. 10, 349–414.

TALLING, J. F. (1961). Photosynthesis under natural conditions, *Annual Review of Plant Physiology,* **12,** 133–154.

WESTLAKE, D. F. (1963). Comparisons of plant productivity, *Biological Reviews,* **38,** 383–425.

Collection of papers

NYBAKKEN, J. W. (1971). *Readings in Marine Ecology.* Sections on *Plankton* (9 papers) and *Concepts in Marine Ecology* (6 papers).

Identification

BOURRELLY, P. Les Algues d'eau douce. (1966 T1). Les Algues vertes; (1968 T2). Les Algues jaunes et braunes; (1970 T3). Eugléniens, Peridiniens, Algues rouges et algues bleues. Boubée: Paris.

BUTCHER, R. W. An Introductory Account of the smaller Algae of British Coastal Waters. *Fishery Investigations,* Series iv. I Introduction and Chlorophyceae (1959); IV Cryptophyceae (1967); VIII Euglenophyceae (1961).

HENDEY, N. I. (1964). An Introductory Account of the smaller Algae of British Coastal Waters. *Fishery Investigations,* Series iv; v Bacillariophyceae.

NEWELL, G. E. and NEWELL, R. (1966). *Marine Plankton: A Practical Guide.* Hutchinson: London.

LEBOUR, M. V. (1930). *The Planktonic Diatoms of Northern Seas.* Ray Society Publication.

LEBOUR, M. V. (1925). *The Dinoflagellates of Northern Seas.* Marine Biological Association of the United Kingdom: Plymouth.

PRESCOTT, G. V. (1954). *How to Know the Freshwater Algae.* Brown: Dubuque, Iowa.

SMITH, G. M. (1950). *Freshwater Algae of the United States.* McGraw Hill: New York.

Measurements

GOLTERMAN, H. L. Ed. (1969). Methods for Chemical Analysis of Fresh Waters. IBP Handbook No. 8 Blackwell: Oxford and Edinburgh.

MACKERETH, F. J. H. (1963). *Some Methods of Water Analysis for Limnologists.* Freshwater Biological Association Publication No. 21, 70 pp.

STRICKLAND, J. D. H. and PARSONS, T. R. (1968). *A Practical Handbook of Seawater Analysis.* Bulletin No. 167 of Fisheries Research Board of Canada: Ottawa.

VOLLENWEIDER, R. A. (1969). *A Manual on Methods for Measuring Primary Production in Aquatic Environments.* IBP Handbook No. 12. Blackwell: Oxford and Edinburgh.

Index